Drywall
Level 1
SECOND EDITION

Pearson **NCCER** | National Center for Construction Education and Research

NCCER
President and Chief Executive Officer: Boyd Worsham
Vice President of Innovation and Advancement: Jennifer Wilkerson
Chief Learning Officer: Lisa Strite
Senior Manager of Projects: Chris Wilson
Project Manager: Patrick Bruce
Testing/Assessment Project Manager: Elizabeth Schlaupitz
Lead Technical Writer: Jo Ann Bartusik
Production Artists: Judd Ivines, Chris Kerston
Permissions Specialist: Adam Black

Pearson
Manager of Project Management: Vanessa Price
Senior Digital Producer: Shannon Stanton
Associate Project Manager: Monica Perez
Content Producer: Alexandrina Wolf
Executive Marketing Manager: Mark Marsden
Designer: Mary Siener
Rights and Permissions: Jenell Forschler, Integra Content Services
Composition: Integra Software Services

Cover image provided by DPR Construction

3 2023

ISBN-13: 978-0-13-817535-1

PREFACE

To the Trainee

Walk into almost any home, apartment complex, or commercial building and look around. The odds are that professional drywall applicators installed the walls and ceilings and placed insulation, soundproofing, and firestopping materials behind and on those walls and ceilings. They may also have applied textures and trims to enhance both the interior and exterior finishes of the buildings.

There were approximately 97,000 drywall and ceiling tile installers working in the United States in 2021. Careers range from installer to specialty finisher to business owner; related professions include sheetrock applicator and acoustical carpenter.

In the first level of drywall training, you will learn the basics including insulation, installation, and finishing. In the second level, you will explore specialized topics such as steel framing, acoustical ceilings, and specialty finishes. In both levels, you will learn about the materials and tools used in the drywall profession.

As you begin drywall training, you take a step toward a satisfying and rewarding career. Continuing your craft education is important as technology and the materials you work with are changing all the time. We wish you success as you begin your construction career in the drywall trade and hope that you will continue your training outside of this series. Taking advantage of the various training opportunities shows your initiative and a desire to learn. The industry's best professionals are equipped with these qualities.

New with *Drywall Level 1*

The second edition of Drywall Level One reflects current tools and practices of the craft. NCCER is proud to release this edition with our latest instructional systems design, linking learning objectives to each module's content. This revised edition expands content around cold-formed steel framing applications to establish more balance between residential and commercial drywall applications.

SCAN ME

We wish you success as you progress through this training program. If you have any comments on how NCCER might improve upon this textbook, please complete the User Update form using the QR code. NCCER appreciates and welcomes customer feedback. You may submit yours by emailing **support@nccer.org**. When doing so, please identify feedback on this title by listing *#DrywallL1* in the subject line.

Our website, **www.nccer.org**, has information on the latest product releases and training.

NCCER Standardized Curricula

NCCER is a not-for-profit 501(c)(3) education foundation established in 1996 by the world's largest and most progressive construction companies and national construction associations. It was founded to address the severe workforce shortage facing the industry and to develop a standardized training process and curricula. Today, NCCER is supported by hundreds of leading construction and maintenance companies, manufacturers, and national associations. The NCCER Standardized Curricula was developed by NCCER in partnership with Pearson, the world's largest educational publisher.

Some features of the NCCER Standardized Curricula are as follows:

- An industry-proven record of success
- Curricula developed by the industry, for the industry
- National standardization providing portability of learned job skills and educational credits
- Compliance with the Office of Apprenticeship requirements for related classroom training (*CFR 29:29*)
- Well-illustrated, up-to-date, and practical information

NCCER maintains a secure online database that provides certificates, digital badges, transcripts, and wallet cards to individuals who successfully complete programs under an NCCER-accredited organization or through one of NCCER's self-paced, online programs. This system also allows individuals and employers to track and verify industry-recognized credentials and certifications in real time.

For information on NCCER's credentials, contact NCCER Customer Service at 1-888-622-3720 or visit **www.nccer.org**.

Digital Credentials

Show off your industry-recognized credentials online with NCCER's digital credentials!

NCCER is now providing online credentials. Transform your knowledge, skills, and achievements into digital credentials that you can share across social media platforms, send to your network, and add to your resume. For more information, visit **www.nccer.org**.

Cover Image

DPR Construction's Self-Perform Drywall group installs drywall across America. The project pictured is a 220,000 square-foot life-science research laboratory and office interior build-out in San Diego, California.

On this project, the Self-Perform Drywall team utilized a robotic field printer that autonomously prints the BIM model onto the floor with 1/16" (1mm) of accuracy. The field printer labels wall and ceiling framing, wall-type ratings, door-rough openings, anchor points, and radius wall layouts. This ultimately ensures that the drywall is installed within the established tolerances.

DESIGN FEATURES

Content is organized and presented in a functional structure that allows trainees to access the information where they need it.

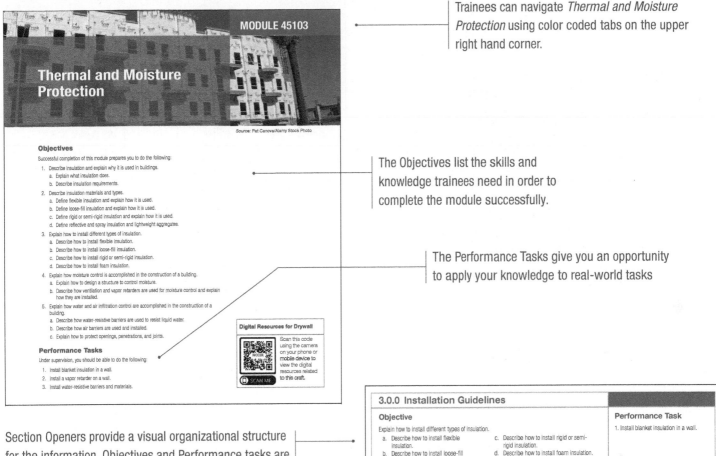

Trainees can navigate *Thermal and Moisture Protection* **using color coded tabs on the upper right hand corner.**

The Objectives list the skills and knowledge trainees need in order to complete the module successfully.

The Performance Tasks give you an opportunity to apply your knowledge to real-world tasks

Section Openers provide a visual organizational structure for the information. Objectives and Performance tasks are broken out for each section.

3.0.0 Installation Guidelines	
Objective	**Performance Task**
Explain how to install different types of insulation. a. Describe how to install flexible insulation. b. Describe how to install loose-fill insulation. c. Describe how to install rigid or semi-rigid insulation. d. Describe how to install foam insulation.	1. Install blanket insulation in a wall.

Trade Terms appear on the page adjacent to the text where they are first presented.

LSL is used for **millwork** such as doors and windows, and any other product that requires high-grade lumber. However, LSL will not support as great a load as a comparable size of PSL because PSL is made from stronger wood.

Wood I-beams consist of a web with flanges bonded to the top and bottom. This arrangement, which mimics the steel I-beam, provides exceptional strength. The web can be made of OSB or plywood. The flanges are grooved to fit over the web. A wood I-beam can be used as a floor joist, **rafter**, or header. Because of its strength, a wood I-beam can be used in greater spans than a comparable length of dimensional lumber.

Glulam is made from several lengths of solid lumber that have been glued together. It is popular in architectural applications where exposed beams are used (*Figure 5*). Because of its exceptional strength and flexibility, glulam can be used in areas subject to high winds or earthquakes.

Millwork: Various types of manufactured wood products such as doors, windows, and moldings.

Rafter: A sloping structural member of a roof frame to which sheathing is attached.

Step-by-step presentations and math equations help make the concepts clear and easy to grasp.

Tape blisters can also happen if the joint is too wide, either because the tape was not properly embedded in the joint compound, or because the tape draws moisture too quickly from the joint compound. Another cause of a blister occurs when topping compound is used instead of joint compound to embed the tape. To repair a tape blister, proceed as follows:

Step 1 Slit the blister with a knife. If the blister is large, cut and remove the entire section of tape that came unbonded.

Step 2 Sand or scrape out enough of the dried joint compound so you can embed a new section of tape.

Step 3 Work joint compound underneath the tape, smoothing the slit in the old or new section of tape into the joint compound as you go. This embeds the blistered section. This is a hand procedure only. Do not attempt to do this with another run of the automatic taper.

Step 4 Apply a skim coat of joint compound over the tape. When this coat is dry, apply the required number of topping coats, always allowing enough drying time in between coats. Sand enough to produce a smooth finish that is flush with the surrounding surface.

QR codes link trainees directly to digital resources that highlight current content.

A material or object's thermal resistance—or ability to resist **heat conduction**—is measured as its **R-value**. As a rule, the higher the R-value, the greater the effectiveness of the insulation. (These requirements are outlined later in the module.) Drywall mechanics must know how to find R-values so they can meet these requirements.

R-value is expressed as:

$$R = 1/k \text{ or } 1/C$$

Where:

k = amount of heat in British thermal units (Btu) transferred in one hour through 1 ft^2 of a material that is 1" thick and has a temperature difference between its surfaces of 1°F; also called the *coefficient of thermal conductivity*

C = conductance of a material, regardless of its thickness; the amount of heat in Btus that will flow through a material in one hour per ft^2 of surface with 1°F of temperature difference

R = thermal resistance; the reciprocal (opposite) of conductivity or conductance

The higher the R-value, the lower the conductive heat transfer. *Table 1* shows the R-values of various common building materials, including some common insulating materials.

Important information is highlighted, illustrated, and presented to facilitate learning.

Placement of images near the text description and details such as callouts and labels help trainees absorb information.

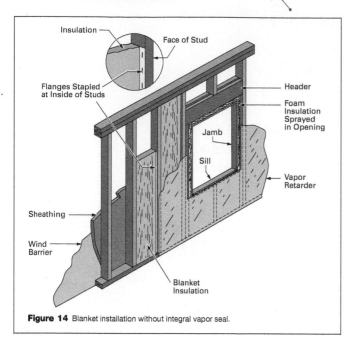

Figure 14 Blanket installation without integral vapor seal.

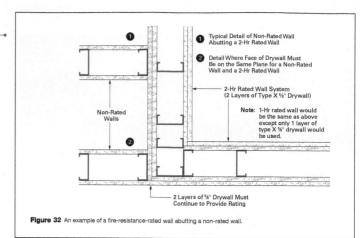

Figure 32 An example of a fire-resistance-rated wall abutting a non-rated wall.

① Typical Detail of Non-Rated Wall Abutting a 2-Hr Rated Wall

② Detail Where Face of Drywall Must Be on the Same Plane for a Non-Rated Wall and a 2-Hr Rated Wall

2-Hr Rated Wall System (2 Layers of Type X ⅝" Drywall)

Note: 1-Hr rated wall would be the same as above except only 1 layer of type X ⅝" drywall would be used.

2 Layers of ⅝" Drywall Must Continue to Provide Rating

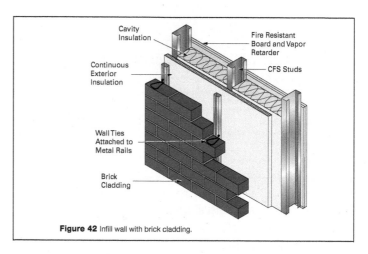

Figure 42 Infill wall with brick cladding.

WARNING!

When working in an area where silica dust is present, use the appropriate respiratory protection. Never work around silica dust without proper training, authorization, and PPE.

CAUTION

Casing and common nails have heads that are too small in relation to the shank; they easily cut into the face paper and threaten the integrity of the board and its attachment to the stud. Nail heads that are too large are also likely to cut the paper surface if the nail is driven incorrectly at a slight angle.

NOTE

Gypsum board measuring ¼" thick is not appropriate for single-ply application on wood or steel framing for ceilings or walls, but it can be used as a base ply or a face ply in multi-ply construction. It can also be applied directly to an existing surface of wood paneling, plaster, concrete, or masonry.

New boxes highlight safety and other important information for trainees. Warning boxes stress potentially dangerous situations, while Caution boxes alert trainees to dangers that may cause damage to equipment. Notes boxes provide additional information on a topic.

Going Green

Recycling Gypsum Board Waste

The gypsum board waste that results from construction projects can be used for many purposes, including as a filler in plastics and cement, and as a soil conditioner that has been shown to reduce the levels of toxic soluble reactive phosphorus in waterways plagued by fertilizer runoff. With the proper separation of waste materials on the worksite—a practice that is growing in popularity—gypsum board without any fasteners or tape can be reclaimed by manufacturers to make new gypsum board. Leading gypsum board brands are now offering panels containing some recycled material. Companies such as USA Gypsum partner with building contractors to recycle gypsum board waste. Since 1998, USA Gypsum has recycled millions of pounds of gypsum board.

Going Green looks at ways to preserve the environment save energy, and make good choices regarding the health of the planet.

Think About It

Fiberglass Insulation

1. Are the flanges on faced fiberglass insulation always stapled to the inside of the stud?
2. Can you increase the effectiveness of fiberglass insulation by squeezing more into a smaller space?

Achieving a Rounded Appearance

A smooth, rounded finish appearance can be obtained by using a bullnose corner molding and cap such as the ones shown here.

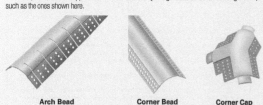

Arch Bead   Corner Bead   Corner Cap

These boxed features provide additional information that enhances the text.

Did You Know?

Mold

Moisture accumulating inside a building can damage the structure and promote the growth of mold. While it is not always harmful, this mold may cause allergic reactions or other respiratory problems in some people. Airborne mold spores can also cause infections, primarily in people whose immune systems are compromised.

Review questions at the end of each section and module allow trainees to measure their progress

3.0.0 Section Review

1. When installing faced insulation in a wall that will have a separate vapor barrier installed, the staples should be fastened to the wall frame _____.
   a. on the faces of the wall studs, top plate, and sole plate
   b. on the inside surfaces of the wall studs, top plate, and sole plate
   c. only to the top plate and sole plate
   d. only to the faces of the wall studs

2. How many markers should be placed in loose-fill insulation per 300 square feet?
   a. 1
   b. 2
   c. 3
   d. 4

3. What is a type of fastener you can use when installing rigid insulation panels to prevent crushing the insulation?
   a. Screws with washers
   b. Nails with small heads
   c. Sheathing tape
   d. Wood glue

4. What should you do before installing foam insulation?
   a. Waterproof the exterior of the foundation.
   b. Install soffit baffles and blocking.
   c. Use plastic to cover everything that shouldn't be touched by foam.
   d. Permanently install strike-off boards.

Module 45103 Review Questions

1. What is the main purpose of insulating a structure?
   a. To control the movement of heat through a wall
   b. To reinforce the structural members with additional stability
   c. To make a structure as soundproof as possible
   d. To finish the interior walls of the structure

2. The two overall categories of insulation are _____.
   a. continuous and rigid
   b. rigid and foam
   c. flexible and loose-fill
   d. cavity and continuous

3. Heat will always flow (or conduct) through any material or gas from _____.
   a. the interior to the exterior
   b. the exterior to the interior
   c. a higher temperature area to a lower temperature area
   d. a lower temperature area to a higher temperature area

4. Which type of insulation is specifically required by the *IECC*® to ensure sufficient insulation?
   a. Fiberglass insulation
   b. Reflective insulation
   c. Continuous insulation
   d. Lightweight aggregates

5. What is one place where insulation should be installed in a building?
   a. Garages
   b. Roofs
   c. In floors above the first floor
   d. In interior walls

ACKNOWLEDGMENTS

This curriculum was revised as a result of the farsightedness and leadership of the following sponsors:

American Wall & Ceiling Institute
Baker Triangle
DPR Construction
FCI Cumberland
Grayhawk, LLC
Marek Brothers
Bill Ford
Johnny Hull
Kevin Howser
Ricardo Menchaca
Ricardo Reyes Aguilar
Richard Bell
Robert Grupe

NCCER PARTNERS

To see a full list of NCCER Partners, please visit **www.nccer.org/about-us/partners**.

SCAN ME

CONTENTS

Module 45102 Construction Materials and Methods

Module 45104 Drywall Installation

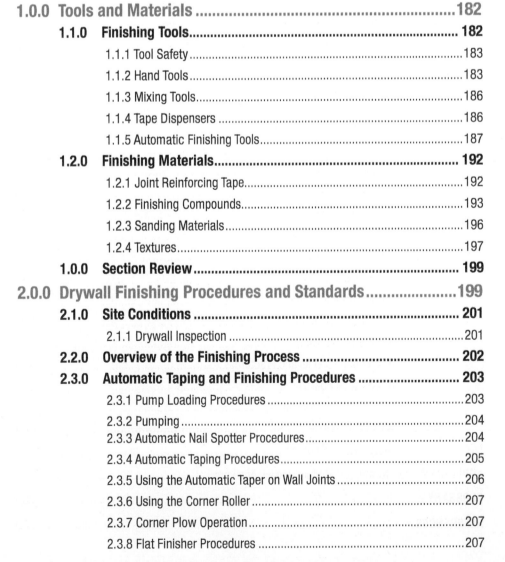

Introduction to Drywall

Source: Viktor Fedorenko/Shutterstock

Objectives

Successful completion of this module prepares you to do the following:

1. Identify safety hazards and precautions associated with the construction industry.
   a. Describe the focus four and explain how to reduce hazards associated with handling drywall materials.
   b. Explain the benefits of a job hazard analysis and how to report unsafe conditions.
2. Describe the purpose and history of the drywall trade.
   a. Identify the processes, materials, and tools needed to install gypsum panels.
3. Identify career and training opportunities in the drywall trade.
   a. Describe craft training opportunities within the drywall trade.
4. Identify skills and attributes of successful drywall mechanics.
   a. List the skills and responsibilities of professional drywall mechanics.
   b. Summarize the traits and standards followed by professional drywall mechanics.

Performance Tasks

This is a knowledge-based module. There are no performance tasks.

Overview

Most homes and businesses use gypsum boards (also known as drywall boards) as the finish for their walls and ceilings. Gypsum board installation and finishing mechanics are needed for just about every building. When drywall professionals finish a job, no matter how many panels are used, walls and ceilings appear as though they are made from a single, smooth sheet. To achieve this level of expertise, drywall mechanics must have a thorough knowledge of specialized tools, materials, and techniques.

NCCER Industry-Recognized Credentials

If you are training through an NCCER-accredited sponsor, you may be eligible for credentials from NCCER. The ID number for this module is 45101. Note that this module may have been used in other NCCER curricula and may apply to other level completions. Contact NCCER at 1.888.622.3720 or go to **www.nccer.org** for more information.

You can also show off your industry-recognized credentials online with NCCER's digital credentials. Transform your knowledge, skills, and achievements into credentials that you can share across social media platforms, send to your network, and add to your resume. For more information, visit **www.nccer.org**.

Digital Resources for Drywall

Scan this code using the camera on your phone or mobile device to view the digital resources related to this craft.

Performance Tasks	1.0.0 Employer and Employee Safety Obligations

1.0.0 Employer and Employee Safety Obligations

Performance Tasks

This is a knowledge-based module. There are no Performance Tasks.

Objective

Identify safety hazards and precautions associated with the construction industry.

a. Describe the focus four and explain how to reduce hazards associated with handling drywall materials.

b. Explain the benefits of a job hazard analysis and how to report unsafe conditions.

When you accept a job as a drywall mechanic, you have a safety obligation to your employer, coworkers, family, and yourself (*Figure 1*). In exchange for your wages and benefits, you agree to work safely. You are also obligated to make sure anyone you work with is working safely. Likewise, your employer is obligated to maintain a safe workplace for all employees. The ultimate responsibility for on-the-job safety, however, rests with you. Whether installing a gypsum panel or unloading and storing building materials and supplies, safety is your responsibility.

Figure 1 Safety is your responsibility.
Source: Poring Studio/Shutterstock

1.1.0 Drywall Safety

Focus four: The four leading causes of death in construction work: falls, struck-by hazards, caught-in or caught-between hazards, and electrical hazards.

The four leading causes of death in construction work are called the **focus four**. The focus four—also known as the *fatal four*—are responsible for almost 60 percent of the construction industry's fatalities. The focus four includes falls, struck-by hazards, caught-in or caught-between hazards, and electrical hazards. As shown in *Figure 2*, the number of deaths from falling exceed the number caused by the other three hazards. While some of these hazards are less likely to happen to drywall mechanics due to the nature of their jobs, you should still take all necessary safety precautions. Always be alert to job conditions that place you or your coworkers in any of these potentially deadly situations.

Explanations of the four leading hazard groups are as follows:

1. Falls from elevation are incidents involving failure to provide, or failure to use appropriate fall protection.

2. Struck-by accidents involve unsafe operation of equipment, machinery, and vehicles, as well as improper handling of materials, such as through unsafe rigging operations.

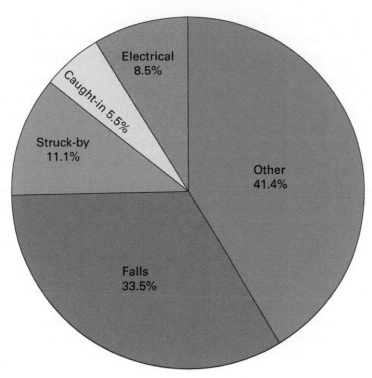

Figure 2 Construction's four leading hazards.
Source: US OSHA, 2018

3. Caught-in or caught-between accidents involve unsafe operation of equipment, machinery, and vehicles, as well as improper safety procedures in a **confined space.**

4. Electrical shock accidents involve contact with overhead wires or use of defective tools.

Confined space: A work area large enough for a person to work in, but with limited means of entry and exit and not designed for continuous occupancy. Crawl spaces and attics are examples of confined spaces.

1.1.1 Personal Protective Equipment

All drywall mechanics are responsible for wearing the appropriate personal protective equipment (PPE) while on the job. When worn correctly, PPE is designed to protect you from injury. It is important to keep PPE in good condition and to know when PPE should be used for a given task. Many workers are injured on the job because they are not using proper PPE. Not all potentially dangerous conditions can be seen just by looking around a jobsite. Stop and consider what type of accidents could occur for any task you are about to perform. Using common sense and knowing how to use PPE greatly reduces your risk of injury, regardless of what construction material is used.

When working with gypsum board, wood, and related tools, it is important to use the following PPE to protect your skin, face, and eyes from dust and debris:

• Face mask
• Safety glasses or goggles
• Hard hat
• Gloves

1.1.2 Good Jobsite Housekeeping

While the bulk of materials on a drywall project are put to use, a certain amount of scrap and other debris is produced during typical processes. Good jobsite housekeeping is required to help prevent accidents. Good housekeeping rules include the following:

• Remove all scrap material from the work area.
• Clean up spills.
• Remove all **combustible** scrap materials regularly.

Combustible: Capable of easily igniting and rapidly burning; used to describe a fuel with a flash point at or above 100°F.

- Make sure there are containers for the collection and separation of refuse. Containers for flammable or harmful refuse must have lids.
- Store all tools and equipment when you are finished using them.

Depending on the size of the project, many types of building materials are delivered to the jobsite in bulk (large quantities). Always use appropriate material-handling equipment when lifting and/or handling heavy objects or large quantities of building materials. When lifting heavy materials, bend your knees, hold the load closely, and keep your back straight while standing up. Many construction injuries can be attributed to improper lifting techniques.

1.1.3 Gypsum Board Safety

Silicosis: A serious lung disease resulting from the inhalation of crystalline silica particles.

Drywall mechanics typically work with gypsum boards, which often contain small amounts of silica. Silica is a mineral found in concrete, masonry, rock, and sand. Silica dust is created when cutting, drilling, sanding, or performing similar tasks on materials containing crystalline silica. Prolonged exposure to silica dust can cause **silicosis**, which is an incurable, and sometimes fatal, lung disease. The time it takes for silicosis to develop varies based on the duration and the amount of silica exposure.

Chronic obstructive pulmonary disease (COPD): Lung diseases that result in the obstruction of lung airflow and interfere with normal breathing. Chronic bronchitis and emphysema generally fall under the diagnosis of COPD.

In addition to silicosis, the inhalation of silica dust increases the risk of lung cancer and **chronic obstructive pulmonary disease (COPD)**. Individuals working closely with materials containing silica also risk the development of kidney disease.

> **WARNING!**
>
> When working in an area where silica dust is present, use the appropriate respiratory protection. Never work around silica dust without proper training, authorization, and PPE.

OSHA holds employers responsible for protecting workers from crystalline silica. *OSHA Standard 29 CFR, Part 1926.1153*, known in the workplace as the *silica standard*, outlines the steps for controlling exposure. Options are available for keeping workers safe, as well as methods to monitor exposure.

The basic requirements for employers include the establishment and implementation of a written silica exposure control plan that identifies tasks leading to exposure and the approaches to protecting workers. In addition, employers must do the following:

- Designate a competent person to ensure the exposure control plan is effectively implemented.
- Establish procedures to restrict access to areas where silica dust is actively generated.
- Train workers on tasks that create silica dust and ways to limit exposure.
- Limit common housekeeping practices that expose workers to silica dust when an alternative is available.
- Provide medical exams that include lung-function tests and x-rays every three years for workers who wear a respirator for 30 or more days each year.
- Keep records of all the above.

Workers should remove any dust from the area frequently using an OSHA-approved HEPA filter vacuum (*Figure 3*).

1.1.4 Wood Product Safety

Figure 3 Vacuum.
Source: SERHII LIAKHEVYCH/
Alamy Stock Photo

One of the most common injuries associated with wood products is embedded splinters. Many of these injuries can be avoided if the proper gloves are worn when handling building materials. Along with preventing embedded splinters, gloves help to prevent cuts or scrapes when handling sharp objects.

Never wear gloves around rotating or moving equipment as they can easily get caught up in the equipment.

When stacking materials such as lumber, ensure they are secure to prevent them from falling or sliding. For lumber, place the pieces flat on the ground, and do not stand them on end. Do not pile lumber more than 6' high if moving it manually.

Chemicals used in treated lumber may present a hazard to people and the environment. Therefore, apply the following precautions when working with treated lumber:

- When cutting treated lumber, always wear eye protection and a dust mask.
- Wash any skin exposed while cutting or handling the lumber.
- Wash clothing exposed to sawdust separately from other clothing.
- Do not burn treated lumber, because the ash poses a health hazard. Check local regulations for proper disposal procedures.
- Be sure to read and follow the manufacturer's safety instructions.

Plywood Safety

Plywood is awkward to carry. Remember, carry only one sheet of plywood at a time, and do not hold it over your head. In strong winds, use caution as the plywood could act like a sail and injure you and others on the jobsite.

1.1.5 Steel Product Safety

Many of the general safety guidelines previously discussed also apply to steel product safety. Additional steel product safety guidelines are as follows:

- Wear gloves when handling steel framing members. Gloves should be thick enough to prevent penetration by sharp edges of the framing members.
- Unlike wood, metal does not absorb moisture. For this reason, steel framing members may become slippery when wet. Use caution when handling wet steel framing members.
- Ensure that proper PPE is used, including hearing protection and goggles, when cutting steel framing members with a cutoff saw. When metal is cut with a cutoff saw, a loud noise may be emitted and flying metal fragments may be produced.
- The edges of steel framing members may be sharp. Avoid dropping members or placing heavy loads of steel framing members on electrical cords as they may cut through the cord and create an electrical hazard.

CAUTION

Some steel framing members are galvanized with a zinc coating to prevent corrosion. When zinc reaches an elevated temperature (over 900°F)—such as when welding or cutting framing members—the zinc burns and emits a zinc oxide vapor that can irritate the lungs and may lead to difficulty breathing. Craftworkers exposed to high concentrations of zinc oxide over prolonged periods of time can develop a condition known as the zinc chills, which can produce fevers and tremors.

1.2.0   Creating a Safety Culture

A safety culture is created when a company sees and embraces the value of a safe work environment. Creating and maintaining a safety culture is an ongoing process that includes sound safety practices and training. Everyone in the company—from management to craftworkers—must be responsible for safety whenever they are on the job. Having a strong safety culture can lower a company's **experience modification rate (EMR)**, which can lead a company to winning more bids and keeping more workers employed. EMR is a number, or rating, assigned to a company based on its history with injuries and its future risks of experiencing more loss. Insurance companies often use this rating to price workers' compensation premiums.

1.2.1 Job Hazard Analysis

Before starting each day, consider the hazards you might encounter and take corrective action to minimize your risk of injury. One way to promote safety in your workplace is to conduct a **job hazard analysis (JHA)** at the start of each

Experience modification rate (EMR): A rating used to determine surcharge or credit to workers' compensation premiums based on a company's accident experience and potential for future losses.

Job hazard analysis (JHA): An approach that emphasizes job tasks to identify hazards before they cause any harm. The focus is on the relationship between the worker, the task, the tools, and the work environment.

Hazard: Something that may be present on the jobsite that can cause immediate harm.

workday. A **hazard** is something that may be present on the jobsite that can cause immediate harm. The JHA is a technique used to assess existing and potential jobsite hazards, understand related risks before they occur, and determine how the risks can be eliminated or controlled. In addition to helping to create a safety culture, the Occupational Safety and Health Administration (OSHA) strongly encourages the use of JHAs on the worksite. Performing regular JHAs has the following key benefits:

- *Recognition of hazardous conditions* — As one of the primary goals and benefits of JHAs, reducing the number of workplace hazards is good for you and your employer. Following OSHA standards is important, and identifying potentially hazardous tasks is an extra benefit of the JHA.
- *Better communication* — JHAs may require input from employees and supervisors working on a project in different areas. The result is increased communication and more opportunity to discover potential safety hazards.
- *Job safety standards compliance* — Along with providing a safer workplace, performing JHAs helps ensure compliance with OSHA standards and provides greater protection from potential legal and financial penalties.
- *Consistency of routines* — JHAs break down each job into steps reviewed by workers. Doing so ensures the routines are fully reviewed and applied consistently on the jobsite.

When performing a JHA, break down tasks into individual steps and then analyze them for potential hazards. If a hazard is identified, certain actions or procedures are recommended to correct or prevent it. *Figure 4* shows an example of a form used to conduct a JHA.

JHAs can also be used as pre-planning tools, which helps ensure that safety is planned into the job. As a drywall mechanic on a jobsite, you may be asked to take part in a JHA during job planning. When JHAs are used as pre-planning tools, they typically review the following information:

- Tools, materials, and equipment needs
- Staffing or manpower requirements
- Duration of the job
- Quality concerns

JOB HAZARD ANALYSIS FORM

Job Title: Date of Analysis:
Job Location: Conducted by:
PPE: Staffing:
Tools, Materials, and Equipment: Duration:

Step	Hazards	Quality Concern	Environmental Concern	New Procedure or Protection

Figure 4 Job hazard analysis form.

Safety Committees

Some employers will have safety committees. If you work for such an employer, you are obligated to that committee to maintain a safe working environment. This means two things:

- Follow the safety committee's rules for proper working procedures and practices.
- Report any unsafe equipment and conditions directly to the committee or your supervisor.

1.2.2 Reporting Unsafe Conditions

Here is a basic rule to follow every working day:

If you see something that is not safe, REPORT IT! Do not ignore it. It will not correct itself. You have an obligation to report it.

Occupational Safety and Health Administration (OSHA) regulations require you to report hazardous conditions. This requirement applies to every part of the construction industry. The easiest way to report unsafe conditions is to tell your supervisor. If that person ignores the unsafe condition, report it to the next highest supervisor. If it is the owner who is being unsafe, let that person know your concerns. If nothing is done about it, report it to OSHA. If you are worried about your job being on the line, think about it in terms of your life, or someone else's, being on the line.

The US Congress passed the *Occupational Safety and Health Act* in 1970. This act also created OSHA. It is part of the US Department of Labor. The job of OSHA is to set occupational safety and health standards for all places of employment, enforce these standards, ensure that employers provide and maintain a safe workplace for all employees, and provide research and educational programs to support safe working practices.

OSHA requires each employer to provide a safe and hazard-free working environment. OSHA also requires that employees comply with OSHA rules and regulations that relate to their conduct on the job. To gain compliance, OSHA can perform spot inspections of jobsites, impose fines for violations, and even stop work from proceeding until the jobsite is safe.

According to OSHA standards, you are entitled to on-the-job safety training. As a new employee, you must be:

- Shown how to do your job safely
- Provided with the required personal protective equipment
- Warned about specific hazards
- Supervised for safety while performing the work

The enforcement of the *Occupational Safety and Health Act* is provided by the federal and state safety inspectors who have the legal authority to make employers pay fines for safety violations. The law allows states to have their own safety regulations and agencies to enforce them, but they must first be approved by the US Secretary of Labor. For states that do not develop such regulations and agencies, federal OSHA standards must be obeyed.

These standards are listed in *OSHA Safety and Health Standards for the Construction Industry (29 CFR, Part 1926)*, sometimes called *OSHA Standards 1926*. Other safety standards that apply to the trade are published in *OSHA Safety and Health Standards for General Industry (29 CFR, Parts 1900 to 1910)*.

The most important general requirements that OSHA places on employers in the construction industry are the following:

- The employer must perform frequent and regular jobsite inspections of equipment.
- The employer must instruct all employees to recognize and avoid unsafe conditions, and to know the regulations that pertain to the job so they may control or eliminate any hazards.
- No one may use any tools, equipment, machines, or materials that do not comply with *OSHA Standards 1926*.
- The employer must ensure that only qualified individuals operate tools, equipment, and machines.

Drugs and Alcohol

When people use drugs and alcohol, they are putting both themselves and the people around them at serious risk. A construction site can be a dangerous environment, and it is important to be alert at all times. The use of drugs and alcohol on the job is an accident waiting to happen. You have an obligation to yourself, your employer, and your fellow employees to work safely. What should you do if you discover someone abusing drugs and/or alcohol at work? If you are taking prescription drugs that might affect your ability to work, you need to inform your supervisor.

1.0.0 Section Review

1. How can silica dust be created?
   a. Cutting wood
   b. Lifting heavy objects
   c. Sanding gypsum board
   d. Drilling into steel

2. Something on the jobsite that may result in an injury or illness is referred to as a(n) _____.
   a. experience modification rate
   b. hazard
   c. accident
   d. job hazard analysis

3. What should you do if you see an unsafe condition?
   a. Consider if reporting it would impact your job.
   b. Report it to your supervisor immediately.
   c. Ignore it if you think it will not affect your work.
   d. Contact OSHA directly before speaking to anyone at the site.

2.0.0 Introduction to Drywall

Performance Tasks	Objective
This is a knowledge-based module. There are no Performance Tasks.	Describe the purpose and history of the drywall trade. a. Identify the processes, materials, and tools needed to install gypsum panels.

Gypsum board: A board with a gypsum core and paper facings. It is a building material generally used for walls and ceilings in residential and commercial buildings. It is also commonly referred to as *drywall*.

Gypsum panel: A panel with a gypsum core and glass mat facings.

Gypsum: A chalky type of rock that serves as the basic ingredient of plaster, gypsum board, and gypsum panel.

Studs: The vertical support members for walls.

Joists: Equally spaced framing members that support floors and ceilings.

Joint: The place where two pieces of material meet. For example, the space between two drywall panels.

Plaster: A compound consisting of lime, sand, and water used to cover walls and ceilings.

Lath: Thin, narrow strips of wood used as a base for plaster.

The **gypsum board** and **gypsum panel** are used to finish walls and ceilings on most residential and commercial buildings. Both products are made of the mineral **gypsum** and are typically manufactured as 4' by 8' panels. These dimensions are compatible with the spacing of the framing to which the panels are fastened. For example, a common spacing for wall **studs** and ceiling **joists** is 16" on center (OC). Gypsum panels are sold in other sizes and can be obtained in a variety of sizes by special order.

Gypsum panels are usually applied directly to wood or steel framing members using screws or nails. Once installed, the **joint** between the panels must be finished with special tapes and finishing compounds so that the surface has a flat, smooth appearance. A drywall mechanic is a person who has learned the specialized techniques needed to properly install and finish gypsum panels. *Figures 5–7* show examples of drywall work.

History of Drywall

By historic standards, the gypsum board is a newcomer to the construction industry. Until the middle of the twentieth century, interior walls were generally finished with **plaster** that was applied over narrow strips of rough wood known as **lath**. One advantage of plaster is that it retarded the spread of fire. The gypsum boards of today are known for the same quality. In fact, gypsum boards have been incorporated into standards that define the construction of fire-resistive walls for residential and commercial construction.

Originally, carpenters, painters, and other trades installed and/or finished gypsum boards. The increased use of the product, combined with the importance of installing and finishing it correctly, led to the emergence of a specialized trade known as drywall mechanics.

Figure 5 Hanging drywall.
Source: Arturs Budkevics/Alamy Stock Photo

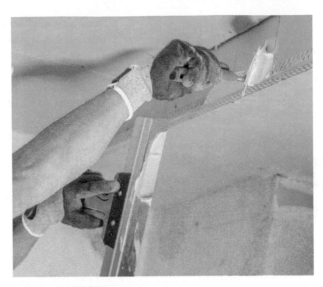

Figure 6 Finishing drywall.
Source: Tomasz Zajda/Alamy Stock Photo

Figure 7 Completed project.
Source: valentyn semenov/Alamy Stock Photo

2.1.0 Modern Drywall Work

Today's gypsum boards and panels are manufactured using an inner core of wet gypsum plaster, along with some additives such as fiberglass and fire retardants. The core is sandwiched between sheets of heavy paper or fiberglass mats and allowed to harden. In the manufacturing process, the gypsum panel is a continuous sheet and is machine-cut to the required length at the end of the process.

The most common type of gypsum panel in use today is $5/8$" Type X fire-resistant panels. These panels can be layered, combined with wood or steel studs, and separated by insulation to achieve specified fire-resistance and sound transmission ratings. Fire-resistance ratings are specified in hours. For example, a wall made with a single layer of $5/8$" Type X attached to wood studs would have a one-hour rating, while two layers of $5/8$" Type X over steel studs would have a two-hour rating. The fire rating defines the time it would take for fire to breach the wall. The rating assumes that any penetrations in the wall, such as those required for piping or wiring runs, have been sealed with a **firestopping** material.

Over time, standardized methods have evolved for placing and fastening gypsum panels. For example, industry experts have determined the correct

Firestopping: A special putty or mechanical device that is used to plug openings in fire-resistive structures such as walls and floors.

placement and spacing for fasteners when installing gypsum panels. Ceiling panels require more fasteners than wall panels. Fewer fasteners are required if screws are used instead of nails.

2.1.1 Tools

A drywall mechanic must learn to use a variety of specialized tools. Gypsum panels are cut to size using a special carbide cutting tool, a utility knife (*Figure 8*), or a gypsum panel saw (*Figure 9*). Other specialized tools such as the power cut-out tool in *Figure 10* are used to cut openings in the panels. For gypsum panel installation, the primary tool is the screw gun (*Figure 11*). The power screwdriver is designed to hold Phillips head screws. A magnetic screwdriver bit is located inside the nose piece. The nose piece is adjustable so that the depth of penetration can be set. The screw gun has a clutch mechanism that disengages the drive when the screw head is below the paper surface of the drywall panel.

Figure 8 Utility knife.
Source: Richard Heyes/Alamy Stock Photo

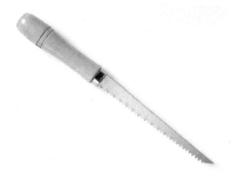

Figure 9 Gypsum panel saw.
Source: Judith Collins/Alamy Stock Photo

Figure 10 Power cut-out tool.
Source: brizmaker/Shutterstock

Figure 11 Screw gun.
Source: videst/123RF

2.1.2 Finishing

Gypsum panel finishing requires a variety of materials and tools. Joints between gypsum panels are finished using a paper or fiberglass mesh tape (*Figure 12*). The tape is usually embedded in the joint with joint compound. This compound, also called mud, comes in powder and premix form (*Figure 13*). A joint typically receives three coats of joint compound, including the tape bedding coat. Each coat is smoothed and sanded.

The joint compound is applied using a taping knife (*Figure 14*). Successive coats are finished with increasingly broader knives.

The Origin of Gypsum Panels

The concept of gypsum panels originally came from a plasterer named Augustine Sackett in the latter part of the nineteenth century. Sackett sandwiched wet plaster of Paris between sheets of heavy paper and allowed it to dry. Although this method achieved some success, it wasn't until after World War II that gypsum panels became popular. Since then, it has become a mainstay of the construction industry.

Figure 12 Gypsum panel joint tape.
Source: Alexander Vinokurov/Alamy Stock Photo

Figure 13 Joint compound.
Source: Richard Levine/Alamy Stock Photo

Figure 14 Taping knives.
Source: Andrii Popov/Alamy Stock Photo

Large Drywall Jobs

For large jobs, some companies prefer to use automatic taping and finishing tools. The automatic taping tool applies tape and joint compound at the same time. The flat finisher is filled with mud and automatically applies and levels the joint compound as it moves along the joint. Other automatic tools are made especially for corner work.

Source: Kevin Howser

Smartphone Apps for Construction

Smartphones are an increasingly popular form of communication that offer a great deal of versatility for drywall mechanics. Smartphone cameras can be used to document on-the-job activities or potential safety violations. Best practices can also be communicated to crew members using video clips. Additionally, construction calculator apps provide craftworkers with the same or even greater capabilities than a handheld calculator.

2.0.0 Section Review

1. How are gypsum board and panels used?
   a. To insulate most residential and commercial buildings
   b. To frame the walls, ceilings, and floors of most residential and commercial buildings
   c. To finish walls and ceilings on most residential and commercial buildings
   d. To provide vertical support beams for walls in most residential and commercial buildings

2. Which type of tool is commonly used to cut gypsum panels?
   a. Circular saw
   b. Screw gun
   c. Finishing knife
   d. Utility knife

3.0.0 Opportunities in the Construction Industry

Objective

Identify career and training opportunities in the drywall trade.
   a. Describe craft training opportunities within the drywall trade.

Performance Tasks

This is a knowledge-based module. There are no Performance Tasks.

Opportunity is driven by knowledge and ability, which are in turn driven by education and training. This NCCER training program was designed and developed by those in the construction industry for the construction industry. It is the only nationally accredited, competency-based construction training program in the United States. A **competency-based training** program requires the trainee to demonstrate the ability to safely perform specific job-related tasks in order to receive credit. This approach is unlike other apprentice programs that merely require a trainee to put in the required number of hours in the classroom and on the job.

The primary goal of NCCER (*Figure 15*) is to standardize construction craft training throughout the country so that both contractors and craftworkers benefit from the training, no matter where they are located. Trainees in an NCCER program receive a certificate for each level of training completed. When you apply for a job with any participating contractor in the country, a copy of your training transcript is available to the employer. If your training is incomplete when making a job transfer, you can pick up where you left off because every participating contractor is using the same training program. Additionally, many technical schools and colleges are using the same program.

The construction industry employs more people and contributes more to the nation's economy than any other industry. Our society will always need new

Competency-based training: Training that places an emphasis on ensuring trainees have the knowledge and skills needed to perform and/or demonstrate specific tasks.

Figure 15 The National Center for Construction Education and Research.

Career: An occupation that offers individuals a lifelong opportunity for training, growth, and advancement.

Apprenticeship: A drywall apprenticeship is focused on gaining on-the-job experience from those who have mastered the craft. Knowledge gained in the classroom is designed to help the apprentice better understand the job's required skills.

homes, highways, bridges and infrastructure, airports, hospitals, schools, factories, and office buildings. These needs create a constant source of good-paying jobs and **career** opportunities for drywall installers and other construction trade professionals. Based on results of the most recent *NCCER Salary Survey*, the average annual salary for drywall installers is approximately $54,773.

As a construction worker, a drywall mechanic can progress from apprentice through the levels of journeyman, master, foreman/lead mechanic/crew leader, and supervisor. Descriptions of each of these levels follow:

- *Journeyman* — After successfully completing an **apprenticeship**, a trainee becomes a journeyman. The term *journeyman* originally meant to journey away from the master and work alone. A person can remain a journeyman or advance in the trade. Journeymen may have additional duties such as supervisor or estimator. With larger companies and on larger jobs, journeymen often become specialists. For example, some drywall mechanics may specialize in installation or finishing.

- *Master* — A master craftworker is one who has achieved and continuously demonstrates the highest skill levels in the trade. A master drywall mechanic is a mentor (guide or coach) for journeymen and apprentices. Master drywall mechanics become supervisors or often start their own businesses.

- *Foreman/lead drywall mechanic/crew leader* — This individual is a frontline leader who directs the work of a crew of drywall mechanics and laborers.

- *Supervisor* — Large construction projects require supervisors who oversee the day-to-day work of crews made up of foremen, journeymen, and apprentices. They are responsible for assigning, directing, and inspecting the work of crew members.

A person working in the drywall trade may work as a drywall installer or a drywall finisher. In addition, the work performed by drywall mechanics varies from company to company and region to region. For example, many drywall contractors perform other work in addition to drywall work. This includes installation of suspended ceilings, steel framing, and thermal insulation (*Figure 16*). Some drywall contractors also install interior doors. Therefore, it's possible that once the building structure is framed, a single contractor will build and finish all the interior walls. In some locations, drywall mechanics install the drywall, but other trades such as painters do the drywall finishing.

Figure 16 Steel framing with thermal insulation.
Source: ronstik/Alamy Stock Photo

The important thing to understand is that a career is a lifelong learning process. To be an effective drywall mechanic, you must keep current with new tools, materials, and methods. If you choose to work into a management role or to someday start a construction business, management and administrative skills will be needed. Every successful manager and business owner started the same way you are starting, and they all have one thing in common: a desire and willingness to continue learning. The learning process begins with apprentice training and must continue throughout your career.

While developing your skills and gaining experience, you will have the opportunity to earn increased pay for your services. The financial incentive for learning and growing within the trade is great. You can't get to the top, however, without learning the basics.

3.1.0 Formal Construction Training

Many skilled drywall mechanics began their formal training in high school while enrolled in building science or construction technology programs. Others were introduced to construction training in college or by their employers. These programs allow students and employees to gain an understanding of basic construction methods and techniques while exploring the field to determine if it is a good fit for them.

The federal government established registered apprenticeship training through the *Code of Federal Regulations (CFR) 29:29*, which dictates specific requirements for apprenticeship, and *CFR 29:30*, which dictates specific guidelines for recruitment, outreach, and registration into approved apprenticeship programs. The US Department of Labor has established specific apprenticeship guidelines, including a minimum number of hours required to complete an apprenticeship.

Education and training throughout the country are undergoing significant change. Educators and researchers have been learning and applying new techniques to adjust to the way students learn and apply their education. New training delivery methods, such as mobile apps, are being developed so students can learn anywhere and at any time.

NCCER is an independent educational foundation founded and funded by the construction industry to provide quality instruction and instructional materials for a wide variety of crafts. One of the ways NCCER accomplishes its mission of creating craft professionals is to offer training and credentialing of the construction workforce. Instead of using the traditional classroom approach to learning, NCCER has adopted a competency-based training method. Competency-based training places emphasis on ensuring the trainee has classroom knowledge *and* the hands-on skills to perform and/or demonstrate specific tasks. NCCER also uses the latest technology, such as interactive computer-based training, to deliver classroom training (*Figure 17*). All completion information for every

Figure 17 Classroom training.

trainee is documented with NCCER, who can then confirm training and skills for workers as they move from company to company, state to state, or even within their own company.

The dramatic shortage of skills within the construction workforce, combined with the shortage of new workers coming into the industry, is providing an opportunity for the construction industry to design and implement new training initiatives. When enrolling in an NCCER program, it is critical that you work for a contractor who supports a national, standardized training program that includes credentials confirming your skill development.

Apprenticeships: Then and Now

The concept of an apprenticeship traces its roots to ancient civilizations in Egypt, Greece, and Rome. Skilled craftworkers employed younger people—usually teenagers—who were unmarried and would live in a craftworker's home. After 7 years with a skilled craftworker, the new "journeyman" would often travel to other towns and villages, continuing to learn from other skilled craftworkers. This approach to being trained by a "master" craftworker is still found in today's apprenticeships. Drywall mechanics gain knowledge of many different trades and skills while developing their craft. This broad set of skills is valuable in the construction industry and will open doors to a wide variety of exciting career opportunities.

3.1.1 Apprenticeship Programs

Apprentice training is a means for individuals entering a craft to learn from those who have already mastered the craft. Its focus is on-the-job performance with others who have worked in the construction field, but it also includes classroom knowledge. Although some theory is presented in the classroom, it is always presented in a way that helps the trainee understand the purpose behind the required skill.

3.1.2 Pre-Apprenticeship Program

A pre-apprenticeship program is also available. It allows students to begin their apprentice training while still in high school. A student entering the program in eleventh grade may complete up to two years of an NCCER program by high school graduation. Additionally, the program, in cooperation with local craft employers, allows students to work in the trade and earn money while still in school. Upon graduation, the student can enter the industry at a higher level and with more pay than someone just starting the apprenticeship program.

This training program is similar to the one used by NCCER, learning centers, contractors, and colleges across the country. Students are recognized through official transcripts and can enter the next year of the program wherever it is offered. They may also have the option of applying the credits at a two- or four-year college that offers degree or certification programs in the construction trades.

3.1.3 Apprenticeship Standards

All apprenticeship standards prescribe certain work-related or on-the-job learning (OJL). This OJL is broken down into specific tasks in which the apprentice receives hands-on training while on the job (*Figure 18*). A specified number of hours are required in each task. The total amount of OJL for a drywall apprenticeship program is traditionally 2,000 hours per year. The program will include two to four years of training (depending on location) and qualifies an individual to become a journeyman drywall mechanic. In a competency-based program, it may also be possible to shorten this time by testing out of specific tasks through a series of performance exams.

The apprentice must log all work time and turn it in to the apprenticeship committee so that accurate time control can be maintained.

Classroom instruction and work-related training do not always run concurrently due to layoffs, availability, or type of work needed to be performed in the field. Furthermore, apprentices with special job experience or coursework may obtain credit toward their classroom requirements. This reduces the total time required in the classroom while maintaining the total OJL requirement. These special cases depend on the type of program and the regulations and standards under which it operates.

Informal OJL provided by employers is usually less thorough than the training provided by a formal apprenticeship program. The degree of training and supervision in an informal program often depends on the size of the employing firm. A small contractor may provide training in only one area, while a large company may be able to provide training in several areas.

Figure 18 On-the-job learning is an important part of an apprenticeship.
Source: RealPeopleGroup/Getty Images

For those entering an apprenticeship program, a high school or technical school education is desirable. Courses in construction/building trades, drafting, and general mathematics are helpful. Manual dexterity, good physical condition, and quick reflexes are important. The ability to solve problems quickly and accurately and to work closely with others is essential. You must also have a high concern for safety.

Ultimately, the prospective apprentice must submit certain information to the Apprenticeship Committee. This information may include the following:

- Aptitude test (General Aptitude Test Battery or GATB Form Test) results, usually administered by the local Employment Security Commission
- Proof of educational background, which is typically a candidate's school transcripts sent to the committee
- Letters of reference from past employers and friends
- Proof of age
- If the candidate is a veteran, a copy of *Form DD214*
- A record of technical training received that relates to the construction industry and/or a record of any pre-apprenticeship training
- High school diploma or General Equivalency Diploma (GED)

The apprentice must be sure to do the following:

- Wear proper safety equipment on the job
- Purchase and maintain tools of the trade, as needed and required by the contractor
- Submit a monthly OJL learning report to the committee
- Report to the committee if a change in employment status occurs
- Attend classroom instruction and adhere to all classroom regulations such as attendance requirements

3.1.4 What to Expect from a Contractor

After an applicant has been selected for apprenticeship by the Apprenticeship Committee, the hiring contractor agrees that the apprentice will be employed under conditions that will result in normal advancement. In return, the contractor requires the apprentice to make satisfactory progress in OJL and related classroom instruction. The contractor agrees that the apprentice will not be employed in a manner that may be considered in violation of the apprenticeship standards. The contractor also agrees to pay a prorated share of the cost of operating the apprenticeship program.

3.1.5 What to Expect from a Training Program

It is important that the contractor you select has a training program. The program should be comprehensive, standardized, and competency-based.

When contractors take the time and initiative to provide quality training, it is a sign that they are willing to invest in their workforce and improve the abilities of their workers. It is important that the training program is national in scope and that transcripts and completion credentials are issued to participants.

Construction is unique in that the contractors share the workforce. A craftworker may work for several contractors throughout their time in the field. In each case, they gain valuable skills in the different types of construction. Therefore, it is critical that the training program help the worker move from company to company, city to city, or state to state without having to start at the beginning for each move. Ask how many contractors in the area use the same program before enrolling. Make sure you always have access to transcripts and certificates to verify your status and level of completion.

Training should also be rewarded. The training program should have a well-defined pay scale attached to it. Successful completion and mastery of skill sets should be accompanied by increases in hourly wages.

Finally, the training curricula should be complete and up to date. Any training program must be committed to maintaining its curricula, developing new delivery mechanisms (Internet, webinars, podcasts, etc.), and teaching new techniques, materials, tools, and equipment in the workplace.

3.1.6 What to Expect from the Apprenticeship Committee

The Apprenticeship Committee is the local administrative body to which the apprentice is assigned, and which oversees the apprentice's training. Every apprenticeship program, whether state or federal, is covered by standards that have been approved by those agencies. The responsibility of enforcement is delegated to the Committee.

The Apprenticeship Committee is responsible not only for enforcement of standards, but also for ensuring that proper training is conducted so that a craftworker successfully completing the program is fully qualified in those areas of training designated by the standards. Among the responsibilities of the Committee are the following:

- Screen and select individuals for apprenticeship and refer them to participating contractors for training.
- Place apprentices under written agreement for participation in the program.
- Establish minimum standards for OJL and related instruction, and monitor the apprentice to see that these criteria are adhered to during the training period.
- Hear all complaints of violations of apprenticeship agreements, whether by contractor or apprentice, and act within the guidelines of the standards.
- Notify the registration agencies of all enrollments, completions, and terminations of apprentices.

3.0.0 Section Review

1. A person who has completed an apprenticeship becomes a _____.
   a. journeyman
   b. master
   c. lead mechanic
   d. supervisor

2. Drywall apprenticeship programs traditionally require a person to complete a minimum of _____ per year of OJL.
   a. 2,000 hours
   b. 4,000 hours
   c. 6,000 hours
   d. 8,000 hours

4.0.0 Drywall Mechanic Skills, Responsibilities, and Characteristics

Performance Tasks

This is a knowledge-based module. There are no Performance Tasks.

Objective

Identify skills and attributes of successful drywall mechanics.
  a. List the skills and responsibilities of professional drywall mechanics.
  b. Summarize the traits and standards followed by professional drywall mechanics.

4.1.0   Skills and Responsibilities

A professional drywall mechanic must have the skills required to use building materials, tools, and equipment to produce a high-quality finished product in a minimum amount of time. Drywall mechanics must be adept at adjusting methods to meet each situation, and they must remain current and knowledgeable about technical advancements in materials, equipment, and craft skills. Additionally, drywall mechanics should never take chances with their own safety or with the safety of others.

4.1.1 Attributes and Skills

Like other building trades, a drywall mechanic's work is active and sometimes strenuous. Prolonged standing, climbing, and squatting are often necessary. Drywall mechanics risk injury from contact with sharp or rough materials and from the use of sharp tools and power equipment. Being new to the trade also increases the chance of being injured. In fact, statistics indicate that most accidents happen within the first 90 days of employment. Remember that the first 90 days on a construction job is a probationary period that allows employers to assess your skills and performance. This is one of the reasons it is essential for you to rely on the knowledge of more experienced workers, learn applicable safety procedures, and wear appropriate PPE.

To be successful in the drywall trade, a person should possess the following attributes:

- Physical strength and ability to lift and move materials
- Hand-eye coordination to use tools
- Ability to communicate clearly with coworkers
- Ability to perform math calculations, to create a **material takeoff**, and to lay out the structure. The takeoff is developed by determining the required materials shown on the project drawings.
- Attention to detail and the ability to measure and cut building materials accurately

Drywall mechanics often have great freedom in planning and performing their work. However, construction skills are standard, and practically all jobs require the skills to perform these basic tasks:

- Gather the materials, tools, and equipment needed for installation and finishing.
- Schedule the work.
- Prepare a job hazard analysis.
- Install gypsum panels using hand and power tools.
- Check the work using levels.

4.1.2 Responsibilities

Drywall mechanics must display a high degree of concern for the safety of workers and for the quality of their work. They must be able to think critically

Material takeoff: A list of building materials obtained by analyzing the project drawings (also known as a *takeoff*).

and evaluate their work at every stage of the project. When they recognize a flaw in the final product, or in the process as something is being built, they take corrective action to resolve or fix the issue. Along with these traits, drywall mechanics have certain responsibilities that must be demonstrated on all projects and jobsites.

Taking Responsibility and Self-Performing

Every drywall mechanic should take responsibility for working safely. Most contractors also expect drywall mechanics to see what needs to be done and then do it. Additionally, when a contractor requests that a specific task be performed, the drywall mechanic should comply promptly. Contractors and foremen should never need to make a request multiple times. Once a responsibility has been delegated, continue to perform the job as directed. Demonstrating these qualities shows a contractor and your coworkers that you are able to perform your job without having to be watched closely.

Following Rules and Regulations

People can work together well only if everyone understands what work is to be completed, how it should be performed, when it should be finished, and who is responsible for completing each task. This is one of the primary reasons rules and regulations are a necessity in any work situation.

Remember that much of a drywall mechanic's work life is governed by the clock. All members of a crew are required to be at work at a specific time. Failure to arrive on time may result in lost time, missed delivery dates, and resentment on the part of those who do come to work on time. Frequent tardiness or absenteeism may also lead to penalties and even dismissal. When accepting a job with a contractor, you agree to the terms of work. Supervisors cannot keep track of people if they arrive any time they please. Ignoring a worker's tardiness is unfair to everyone on the job, especially because failure to be on time may hold up the work of other craftworkers. Better planning of your morning routine will often keep you from being delayed and prevent a late arrival. In fact, arriving a little early indicates your interest in and enthusiasm for your work, which is appreciated by contractors. The habits of being late or missing work are factors that can also prevent or delay your promotion.

It is sometimes necessary to take time off from work. No one should be expected to work when sick or when there is a serious issue to address. However, it is possible to get into the habit of letting unimportant and unnecessary matters keep you from the job. This results in lost production and hardship on those forced to work with less help. The contractor who hires you has a right to expect you to be on the job unless there is a very good reason for staying away.

If it is necessary to stay home, then contact the office early so your supervisor can find someone to replace you for the day. Being courteous means that you let your supervisor know if you are unable to make it to work.

The most frequent causes of absenteeism are illness, death in the family, accidents, personal business, and dissatisfaction with the job. Some of the causes are legitimate and unavoidable, while others can be controlled or avoided.

Contractors sometimes resort to docking pay, demotion, and even dismissal to control tardiness and absenteeism. No contractor likes to impose these types of penalties, but in fairness to workers who do arrive on time and who do not miss work, a contractor may be forced to impose tougher consequences.

Avoiding Tardiness and Absenteeism

Tardiness means you show up late at the jobsite, and absenteeism means you do not show up at all. Consistent tardiness and frequent absences are an indication of poor work habits, unprofessional conduct, and a lack of commitment to your contractor.

The Customer

When you are on a jobsite, consider yourself to be working for both your contractor and the customer. If you are honest and maintain a professional attitude when interacting with customers, everyone will benefit. Your contractor will be pleased with your performance, and the customer will be happy with the work being performed. Try seeing things from a customer's point of view and remember that a good, professional attitude goes a long way toward ensuring repeat business.

Late for Work

Showing up on time is a basic requirement for just about every job. Your contractor is counting on you to be there at a set time, ready to work. While legitimate emergencies may arise that can cause you to be late for or even miss work, consistent tardiness is a bad habit. What are the possible consequences that you could face as a result of tardiness and absenteeism?

4.2.0 Professional Standards

To be successful on the job, a drywall mechanic must not only perform the responsibilities assigned by a supervisor, but also demonstrate strong personal characteristics. These characteristics include the following:

1. Professionalism
2. Honesty
3. Loyalty
4. Willingness to learn
5. Willingness to cooperate
6. Commitment to quality and safety
7. Positive attitude

4.2.1 Professionalism

The word *professionalism* is a broad term that describes the desired overall behavior and attitude expected in the workplace. Many people believe professionalism must be displayed only by those in management. While it is true that management should support and display professionalism, it is also important for drywall mechanics to recognize the role professionalism plays in every aspect of their own job.

Professionalism can be demonstrated in a variety of ways while on the jobsite. From the way you treat your coworkers to the way you communicate with your supervisor, you should display professionalism at all times while on the job. Likewise, you should never tolerate unprofessional behavior of coworkers. This is not to say that you should avoid the unprofessional worker, but you should work to demonstrate to your coworker the benefits of professional behavior. Ultimately, professionalism benefits both the contractor and the worker. It is a personal responsibility, and it is one of the ways you show your commitment to your job and to the drywall trade.

4.2.2 Honesty

Honesty and personal integrity are important characteristics of a successful drywall mechanic. Professionals pride themselves on performing a job well while being punctual and dependable. Each job is completed in a thorough and professional way, not by cutting corners or by reducing materials. A valued professional maintains work attitudes and ethics that protect property such as tools and other materials belonging to contractors, customers, and different craftworkers on the jobsite.

Honesty and success go hand in hand. Your choice is not simply between good or bad, but between success or failure. Dishonesty will always catch up with you. Whether a person is stealing materials, tools, and equipment, or lying about their work, the supervisor will eventually find out. Of course, a drywall mechanic can always find another contractor, but this option will ultimately run out when the drywall mechanic's dishonest reputation becomes well known.

Honesty is one of the keys to ensuring you are in demand as a drywall mechanic and that you bring home a consistent paycheck. Be sure to demonstrate honesty in every jobsite situation and you will reap a variety of rewards. Being honest means more than giving a fair day's work for a fair day's pay. It means always doing what you say you will do, and always telling the truth. Over time, your reputation as an honest, quality-driven drywall mechanic will provide you with many opportunities, and it will establish you as a skilled craftworker who can be trusted on any jobsite.

4.2.3 Loyalty

Craftworkers expect contractors to look out for their interests, to provide them with steady employment, and to promote them to better jobs as their skills increase and jobs open up. Contractors expect their workers to be loyal and to speak well of the company and their coworkers. They also expect craftworkers to be careful with jobsite information, and to keep confidential all matters that pertain to the business or project. Both contractors and workers should remember, however, that loyalty must be displayed by both parties, and it should be consistent over time. By doing so, both loyalty and trust will grow, and the relationship between worker and contractor will be strengthened.

4.2.4 Willingness to Learn

Every contractor has a unique way of doing things. Contractors expect their workers to be willing to learn these ways. The ability to adapt to change quickly and to be willing to learn new methods and techniques is important. Sometimes the purchase of new tools or equipment requires experienced drywall mechanics to learn new methods and operations. In these cases, some craftworkers may resent having to learn new tools and techniques, especially when retraining is required. Try to remember that your construction techniques must be efficient, and your contractors must continue to make a profit. Without improved efficiencies and profit, contractors may be forced to reduce the size of their crews, which may ultimately impact your own job.

Ethical Principles for Members of the Construction Trades

Honesty: Be truthful in all dealings. Conduct business according to the highest professional standards. Faithfully fulfill all contracts and commitments. Do not deliberately mislead or deceive others.

Integrity: Demonstrate personal integrity and the courage of your convictions by doing what is right even under pressure to do otherwise. Do not sacrifice your principles for expediency, be hypocritical, or act in an unscrupulous manner.

Loyalty: Be worthy of trust. Demonstrate fidelity and loyalty to companies, contractors, fellow craftworkers, and trade institutions and organizations.

Fairness: Be fair and just in all dealings. Do not take undue advantage of another's mistakes or difficulties. Fair people display a commitment to justice, equal treatment of individuals, tolerance for and acceptance of diversity, and open-mindedness.

Respect for others: Be courteous and treat all people with equal respect and dignity regardless of sex, race, or national origin.

Respect for the law: Abide by laws, rules, and regulations relating to all personal and business activities.

Commitment to excellence: Pursue excellence in performing your duties, be well informed and prepared, and constantly endeavor to increase your proficiency by gaining new skills and knowledge.

Leadership: By your own conduct, seek to be a positive role model for others.

4.2.5 Willingness to Cooperate

To cooperate means to work together. In our modern business world, cooperation and clear communication are keys to getting things accomplished. Learn to work as a member of a team with your contractor, supervisor, and fellow workers (*Figure 19*) so that you can get your work done efficiently, safely, and on time. Do not underestimate the importance of cooperating with others.

The term *human relations* is often associated with the willingness to cooperate. While being friendly, pleasant, courteous, cooperative, and adaptable are all related to human relations, it is much more than just getting people to like you. It involves knowing how to handle difficult situations as they arise, and maintaining relationships with others with whom you may disagree.

Workers with good human relations skills know how to work with supervisors who are often demanding and may at times seem unfair. They understand the personality traits of others as well as themselves. They build sound working relationships even in difficult situations. They also know how to restore working relationships that have deteriorated and how to handle frustration without taking it out on others.

Remember that effective human relations is directly related to productivity and that productivity is a key factor in a drywall mechanic's success. Every craftworker is expected to produce at a certain level. Contractors quickly lose interest in a worker who may have a great attitude but is unable to keep up with the schedule. To be productive, do your share—or more than your share—without antagonizing your fellow workers. Perform your duties as a drywall mechanic in a manner that encourages others to follow your example.

Figure 19 Cooperation and communication are keys to effective teamwork.
Source: RealPeopleGroup/Getty Images

4.2.6 Commitment to Quality and Safety

All drywall mechanics must be committed to producing quality work. In fact, both novice and experienced drywall mechanics are measured by the quality of their work. Poor quality impacts your reputation and the reputation of your employer. It is also expensive because it costs extra money to fix problems caused by poor workmanship. These types of repairs are called *rework*, and recent studies have shown that contractors typically end up paying five percent or more of the value of a contract because of rework costs. Rework can be the result of poor quality. It can also be caused by something as simple as not paying attention to the order of construction activities.

While one measure of quality is whether the final product is built to specifications and functions as intended, another related measure that drywall mechanics should always consider is safety. Like rework, failure to perform your job safely can result in significant costs for a contractor, and it may even result in bodily injury or death for the careless craftworker.

Working with Other Trades

Cooperation among the various trades at the jobsite and respect for the work of other trades are essential to achieve a smooth-running project. On many well-run jobs, a sense of togetherness (cooperation) develops and the trades work in harmony. Many times, there is even a trade-off of activities that allows the project to progress at a uniform pace. Cooperation makes work flow better and helps ensure timely completion of the project.

4.2.7 Positive Attitude

A positive attitude is essential to a successful construction career. Being positive means being energetic, highly motivated, attentive to others, and always aware of your surroundings. A positive attitude is essential to safety on the job and often contributes to increased productivity of others around you. There may be times when you display a negative attitude at work, but remember that a persistently negative attitude can spoil the positive attitudes of others, and it may make others reluctant to work with you. People favor a positive person. Being positive makes a person's job more interesting, and it can play a big role in your future success with a contractor. Supervisors judge workers' attitudes by the way they approach their jobs, by their reaction to directives, and by the way they handle challenges.

A positive attitude is far more than a smile. In fact, some people transmit a positive attitude even though they seldom smile. They do this by the way they treat others, by the way they look at their responsibilities, and by the way they approach jobsite challenges and difficulties.

The following suggestions can help you maintain a positive attitude:

- Remember that your attitude follows you wherever you go. If you make a greater effort to be a more positive person in your personal life, it will automatically help you on the job.

- Negative comments are seldom welcomed by fellow workers on the job. Neither are they welcomed outside the job. The solution: talk about positive things and be complimentary. Constant complainers do not build healthy and fulfilling relationships—or jobs.

- Look for the good things in people on the job, especially your supervisor and coworkers. Nobody is perfect, but almost everyone has a few worthwhile qualities. If you dwell on a worker's good qualities, it will be easier to work with them.

- Look for the good things where you work. What are the factors that make it a good place to work? Is it the hours, the physical environment, the people, the actual work being done, or is it the atmosphere? Keep in mind that you cannot be expected to like everything. No work assignment is perfect, but if you concentrate on the good things, the negative factors will seem less important and bothersome.

- Look for the good things in the contractor. Almost all organizations have some good qualities. Do not expect to have everything you would like, but there should be enough to keep you positive. In fact, if you decide to stick with a contractor for a long period of time, it is wise to look at the good features and think about them. If you think positively, you will act the same way.

- You may not be able to change the negative attitude of another worker, but you can control and protect your own attitude.

Teamwork

Many of us like to follow all sorts of different teams: racing teams, baseball teams, football teams, and soccer teams. Like sports teams, everyone on a jobsite is part of a team. As a part of that team, you have a responsibility to your teammates. What does teamwork really mean on the job? Craftworkers must sincerely do everything they can to build strong, professional working relationships with fellow craftworkers, supervisors, and customers.

4.0.0 Section Review

1. Statistics indicate that most injuries occur during the first _____ days on the job.
   a. 30
   b. 90
   c. 45
   d. 60

2. A broad term that describes the desired overall behavior and attitude expected in the construction workplace is _____.
   a. closed-mindedness
   b. professionalism
   c. cooperation
   d. strategy

1. What is the deadliest construction job hazard?
   a. Falling
   b. Getting caught between objects
   c. Electrocution
   d. Getting struck by objects

2. During a job hazard analysis, tasks are broken down into _____.
   a. steps
   b. phases
   c. chapters
   d. codes

3. If you see an unsafe condition on the job, you should _____.
   a. ignore it because it is not your job
   b. tell a coworker
   c. call OSHA
   d. report it to a supervisor

4. The purpose of OSHA is to _____.
   a. catch people breaking safety regulations
   b. make rules and regulations governing all aspects of construction projects
   c. ensure that the employer provides and maintains a safe workplace
   d. assign a safety inspector to every project

5. When should an unsafe condition be reported to OSHA?
   a. As soon as it is discovered
   b. Never
   c. If company leaders ignore your report
   d. After it causes an injury

6. The common dimensions of a drywall panel are _____.
   a. 3' × 3'
   b. 4' × 8'
   c. 5' × 8'
   d. 5' × 10'

7. How are gypsum panels usually applied to wood or steel framing members?
   a. Wood glue
   b. Crown staples
   c. Screws or nails
   d. Wedge anchor

8. A common spacing for wall studs and ceiling joists is _____ on center.
   a. 12"
   b. 16"
   c. 18"
   d. 20"

9. A wall made with a single layer of _____ Type X gypsum board attached to wood studs would have a one-hour fire-resistance rating.
   a. $\frac{1}{2}$"
   b. $\frac{3}{8}$"
   c. $\frac{5}{8}$"
   d. $\frac{1}{4}$"

10. Specialized tools, such as the _____, are used to cut openings in gypsum board.
    a. utility knife
    b. power cut-out tool
    c. drywall saw
    d. drywall hatchet

11. What is the first step in finishing gypsum panels?
    a. Screw each panel into studs on the walls and ceiling
    b. Embed joint tape into the joint with a joint compound
    c. Apply two additional coats of mud to the panel
    d. Smooth and sand each coat of mud

12. The _____ is most likely to handle the day-to-day operations on the jobsite.
    a. supervisor
    b. project manager
    c. architect
    d. owner

13. The *Code of Federal Regulations (CFR) 29:30* specifies requirements for _____.
    a. on-the-job safety classes
    b. supervisory training
    c. fall prevention procedures
    d. approved apprenticeship programs

14. Apprenticeship training focuses on classroom knowledge and _____.
    a. on-the-job training
    b. previous skills
    c. length of employment
    d. soft skills

15. The purpose of the pre-apprenticeship program is to _____.
    a. make sure all young people know how to use basic construction tools
    b. provide job opportunities for people who quit high school
    c. allow students to start in an apprenticeship program while still in high school
    d. make sure that people under 18 have proper supervision on the job

16. The group responsible for enforcing apprenticeship standards and ensuring the apprentice is properly trained is the _____.
    a. contractor
    b. Occupational Safety and Health Administration
    c. Department of Commerce
    d. Apprenticeship Committee

17. If you must miss a day of work, you should _____.
    a. make up a good excuse
    b. call in early in the morning to tell your supervisor you won't be there
    c. ask a coworker to let the supervisor know you'll be out
    d. deal with the problem tomorrow

18. The characteristic of _____ means always doing what you say you will do, and always telling the truth.
    a. honesty
    b. willingness to learn
    c. professionalism
    d. loyalty

19. Workers with good _____ skills understand the personality traits of others as well as themselves.
    a. human relations
    b. positive attitude
    c. time management
    d. communication

20. If one of your coworkers complains about your company, you should _____.
    a. contribute your own complaints to the conversation
    b. agree with the person to avoid conflict
    c. suggest that the person look for another job
    d. find some good things to say about the company

Answers to Odd-Numbered Module Review Questions are found in *Appendix A.*

Answers to Section Review Questions

Answer	Section	Objective
Section One		
1. c	1.1.3	1a
2. b	1.2.1	1b
3. b	1.2.2	1b
Section Two		
1. c	2.0.0	2a
2. d	2.1.1	2a
Section Three		
1. a	3.0.0	3a
2. a	3.1.3	3a
Section Four		
1. b	4.1.1	4a
2. b	4.2.1	4b

Construction Materials and Methods

Objectives

Successful completion of this module prepares you to do the following:

1. Describe construction projects, scheduling, and materials used for construction.
   a. List and describe types of lumber, plywood, and building boards.
   b. Describe gypsum board, gypsum products, and MgO boards.
   c. List and describe types of masonry materials.
   d. List and describe types of metal materials.
2. Explain wall construction processes.
   a. Describe wood frame wall construction.
   b. Describe steel frame wall construction.
   c. Describe differences in commercial and residential walls.
3. Explain ceiling construction processes.
   a. Describe residential ceiling construction.
   b. Describe commercial ceiling construction.
4. Describe fire-rated and sound-rated construction and explain why they are important.
   a. Identify fire and sound rating requirements for walls.
   b. Define firestopping and explain how it is accomplished.
   c. Describe sound isolation construction and requirements.

Performance Tasks

This is a knowledge-based module. There are no Performance Tasks.

Overview

Whether you are installing gypsum board or suspended ceiling grids, it is essential to be familiar with the construction methods used in the industry and the materials used in the various types of structures. This module introduces types of structures and building materials drywall mechanics are likely to encounter.

Digital Resources for Drywall

Scan this code using the camera on your phone or mobile device to view the digital resources related to this craft.

1.0.0 Basic Construction and Building Materials

Performance Tasks

This is a knowledge-based module. There are no Performance Tasks.

Objective

Describe construction projects, scheduling, and materials used for construction.

a. List and describe types of lumber, plywood, and building boards.
b. Describe gypsum board, gypsum products, and MgO boards.
c. List and describe types of masonry materials.
d. List and describe types of metal materials.

Each day, workers on a construction site need to install or work with a variety of different materials. Even if you do not need to install a specific material, it is important to know what that material is and how it is used. You may need to treat it a certain way when installing your part of the project. For instance, you must know how to attach drywall to the type of wood or CFS stud being used in the building.

A construction project requires a lot of planning and scheduling because different trades, equipment, and materials are needed at different times in the process. Drywall is normally installed after the building has been dried in (i.e., the exterior siding and roofing are applied so the building remains dry, but the framing is exposed on the inside of the building).

Project planning and scheduling will be covered in more detail later in your training. For now, *Figure 1* and *Figure 2* provide an overview of where each trade fits into the construction process for residential and commercial projects, respectively.

Residential and commercial construction methods are very different. The term *residential* refers to structures that are used for private living. These include single-family and multi-family dwellings, such as houses, townhouses, and apartment buildings. The term *commercial* refers to a building where commerce, or the exchange of goods or services, takes place. Keep in mind, however, that some small commercial buildings may use the same construction techniques

> **NOTE**
>
> The *International Building Code*® (*IBC*®), *International Residential Code*® (*IRC*®), and National Fire Protection Agency® (*NFPA*®) each has codes, requirements, and regulations for building construction. There may also be local codes that apply to a project. Be sure to check applicable codes for all building projects.

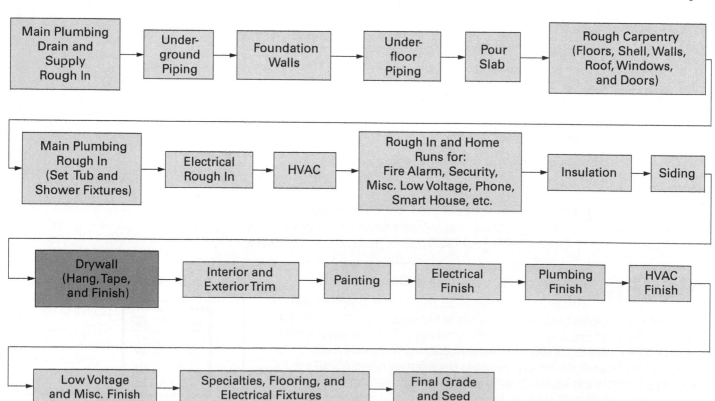

Figure 1 Typical residential construction schedule.

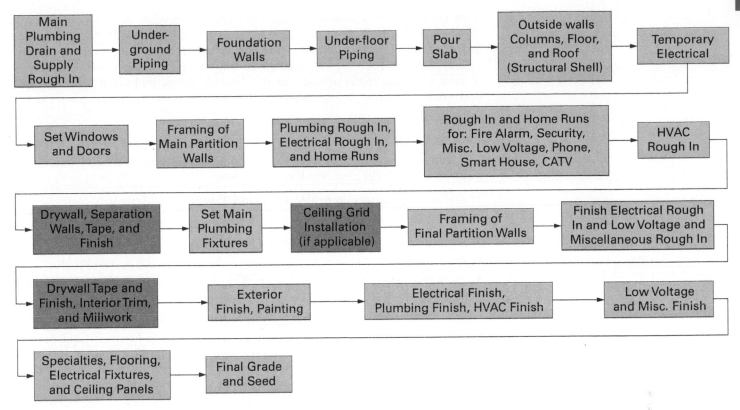

Figure 2 Typical commercial construction schedule.

and materials as those used in residential buildings. High-rise and large apartment buildings can have the same construction requirements as commercial buildings. It's always important to check applicable codes for the correct categorization and requirements for any building project.

Many different materials are used in the construction of a building. Wood or metal frame construction is most common in residential work. Both types must be load-bearing methods where the primary structure is a framed system.

The construction of large commercial buildings such as office buildings, warehouses, apartment buildings, and parking garages generally involves the use of hot rolled steel, cast-in-place concrete, or CMU with non-structural curtain wall exterior steel stud walls and light gauge steel-framed interior walls.

Basic construction materials include various types of lumber, building boards, gypsum boards, MgO boards, steel, and masonry materials. Drywall mechanics must be familiar with these materials and their uses.

Drywall mechanics will also need to be familiar with different types of insulation materials, as well as materials for installing and finishing drywall (caulk, joint compound, paper tape, etc.). These will be discussed in the next modules.

1.1.0 Lumber and Plywood

Lumber is one of the main materials used in US construction. It has been used for thousands of years to build structures all over the world. These days, it is primarily used in small single-family residential and mid-rise framing.

The United States is one of the only places in the world where wood is still the dominant material for new homes. The popularity of using wood globally has gone down over the years for several reasons. For one, wood is a resource that takes time to grow and develop. It is expensive and time-consuming to manufacture, and cutting down forests adds to pollution and climate change. Wood is also highly flammable and buildings that are made of wood are vulnerable to increasingly severe forest fires.

There are more sustainable and fire-resistant alternatives to lumber. Steel and concrete are both more fire-resistant materials. While manufacturing steel does contribute to emissions, most steel can be recycled after use, making it more eco-friendly. In the United States, steel framing is rapidly rising in popularity for both commercial and residential structures.

It is important for drywall mechanics to know how to work with every type of core building material used in the United States, including lumber, steel, and concrete, while being aware of the impact these materials have on the environment.

1.1.1 Lumber

The framework of a single-family or two-family dwelling is usually built from lumber, which is divided into five categories:

- *Boards* — Members up to $1\frac{1}{2}$" thick and 2" wide or wider
- *Light framing (LF)* — Members 2" to 4" thick and 2" to 4" wide
- *Joists and planks (J&P)* — Members 2" to 4" thick and 6" wide or wider
- *Beams and* **stringers** *(B&S)* — Members 5" and thicker by 8" and wider
- *Posts and timbers (P&T)* — Members 5" × 5" and greater, and approximately square

Nearly all lumber used in framing a house is softwood such as pine or fir. Hardwoods such as oak and maple are used primarily in furniture and decorative pieces. Light framing lumber, studs, joists, and planks are all classified as **dimensional lumber**.

You are probably familiar with the terms 2 × 4, 1 × 6, and so on. These numbers represent the nominal (rough) size of the lumber in inches. Once the lumber is dressed (finished) at the lumber mill, it is somewhat smaller, typically $\frac{1}{2}$" to $\frac{3}{4}$" less than the nominal size in each dimension. *Table 1* shows the final dimensions for some standard sizes of softwood dimensional lumber. Note that these dressed dimensions apply only to softwoods; hardwoods have different conversion tables.

Stringers: The support members at the sides of a staircase; also, the timbers used to support formwork for a concrete floor.

Dimensional lumber: Any lumber within a range of 2" to 5" thick and up to 12" wide.

Sill plate: A horizontal timber that supports the framework of a building. It forms the transition between the foundation and the frame.

TABLE 1 Nominal and Dressed Sizes of Dimensional Lumber (in inches)

Nominal	Dressed
1 × 2	$\frac{3}{4} \times 1\frac{1}{2}$
1 × 4	$\frac{3}{4} \times 3\frac{1}{2}$
1 × 6	$\frac{3}{4} \times 5\frac{1}{2}$
1 × 8	$\frac{3}{4} \times 7\frac{1}{2}$
2 × 2	$1\frac{1}{2} \times 1\frac{1}{2}$
2 × 3	$1\frac{1}{2} \times 2\frac{1}{2}$
2 × 4	$1\frac{1}{2} \times 3\frac{1}{2}$
2 × 6	$1\frac{1}{2} \times 5\frac{1}{2}$
2 × 8	$1\frac{1}{2} \times 7\frac{1}{4}$
2 × 10	$1\frac{1}{2} \times 9\frac{1}{4}$
2 × 12	$1\frac{1}{2} \times 11\frac{1}{4}$
4 × 4	$3\frac{1}{2} \times 3\frac{1}{2}$
4 × 6	$3\frac{1}{2} \times 5\frac{1}{2}$
4 × 8	$3\frac{1}{2} \times 7\frac{1}{4}$
4 × 10	$3\frac{1}{2} \times 9\frac{1}{4}$
6 × 6	$5\frac{1}{2} \times 5\frac{1}{2}$

Ghost Wood

Longleaf pine and bald cypress are praised by carpenters for their outstanding beauty and durability. Unfortunately, clear-cutting that occurred in the late 1800s wiped out the two species from the vast forests in the southern United States. Today, lumber companies have resurrected this wood by retracing the rivers that timber companies used to transport the logs to sawmills. Logs of original-growth pine and cypress have been lying along the bottom of these riverbeds for more than one hundred years. What was left behind is now harvested by scuba divers, then sawed, dried, and used for various construction purposes.

Pressure-Treated Lumber

Pressure-treated lumber is softwood lumber protected by chemical preservatives forced deep into the wood through a vacuum-pressure process. It has been used for many years in on-ground and below-ground applications. For example, a landscape timber, a **sill plate**, and a building foundation can all be made of pressure-treated lumber. In some parts of the country, it is also used extensively in the building of decks, porches, docks, and other outdoor structures. It is popular for these uses in areas where structures are exposed to snow for several months of the year. A major advantage of pressure-treated lumber is its relatively low price in comparison with redwood and cedar. When natural woods such as these are used, only the more expensive heartwood will resist decay and insects.

Because the chemicals used in pressure-treated lumber present some hazards to people and the environment, the following special precautions apply to its use:

- When cutting pressure-treated lumber, always wear eye protection and a dust mask.
- Wash any skin that is exposed while cutting or handling the lumber.
- Wash clothing that is exposed to sawdust separately from other clothing.

- Do not burn pressure-treated lumber, as the ash poses a health hazard. Bury it or put it in the trash.
- Be sure to read and follow the manufacturer's safety instructions as defined in the safety data sheet (SDS).

One place to look for pressure-treated lumber is any location where wood comes into contact with the ground, or outdoors where the wood is exposed to moisture.

1.1.2 Plywood

Another commonly used type of building material is plywood.

Plywood is made by gluing together thin layers of wood known as *veneers*. Plywood can have three or more plies (layers). These are bonded together at right angles with glue and heat under tremendous pressure. Putting the plies together at right angles increases the strength; also, the more plies there are, the greater the strength. The ply that is in the center is called the *core* and each of the exposed plies is called a **veneer** or *face* (*Figure 3*). All other plies between the core and veneer are called *crossbands*. Constructing the plywood with the grain of adjacent plies running at right angles reduces the possibility of warping.

Plywood panel thickness varies from $3/16$" to $1\frac{1}{4}$". Common sizes are $\frac{1}{4}$", $\frac{1}{2}$", and $\frac{5}{8}$" for finish paneling and $\frac{3}{8}$", $\frac{1}{2}$", $\frac{5}{8}$", and $\frac{3}{4}$" for some structural purposes. The three types of edges on plywood are butt joint (two standard pieces joined), **shiplap** cut or edge, and interlocking tongue-and-groove. Opposite edges or all four edges may be cut to match.

Plywood panels generally come in standard sizes of 4' × 8', 4' × 9', and 4' × 4'. A few companies produce plywood from 6' to 8' widths and up to 16' in length. **Sheathing**-grade plywood is nominally sized by the manufacturer to allow for expansion; for example, 4' × 8' is really $47\frac{3}{4}$" × $95\frac{3}{4}$".

Plywood is rated by the American Plywood Association (APA) for interior or exterior use. Exterior-rated plywood is used for sheathing, siding, and other applications where there may be exposure to moisture or wet weather conditions. Exterior plywood panels are made of high-grade veneers bonded together with a waterproof glue that is as strong as the wood itself. CDX is one type of veneer plywood with C and D grades on either side that can be used for subflooring, roof sheathing, and exterior sheathing. **Oriented strand board (OSB)** is a less expensive version of CDX that can also be used for roof sheathing or as the substrate under decking or shingles.

Plywood: A building material made of thin layers of wood that is used for sheathing and siding.

Veneer: The covering layer of material for a wall or the facing materials applied to a substrate.

Shiplap: Lumber with edges that are shaped to overlap adjoining pieces.

Sheathing: The sheet material or boards used to close in walls and roofs.

Oriented strand board (OSB): Panels made from layers of wood strands bonded together.

Figure 3 Plywood.
Source: jocic/Shutterstock

Oriented Strand Board (OSB)

Oriented strand board (OSB) is a manufactured structural panel used for wall and roof sheathing and single-layer floor construction. OSB consists of compressed wood strands arranged in three perpendicular layers and bonded with phenolic resin. Some of the qualities of OSB are dimensional stability, stiffness, fastener holding capacity, and no voids in the core material. Before cutting into OSB, be sure to check the applicable manufacturer's safety data sheet (SDS) for safety hazards. The SDS is the most reliable source of safety information.

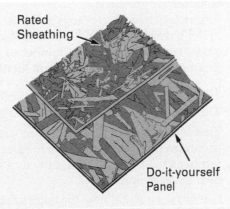

Rated Sheathing

Do-it-yourself Panel

Selecting Plywood

Plywood that is expressly manufactured for either interior or exterior use may be used for other purposes in certain situations. Some local codes may require the use of pressure-treated plywood for exterior construction, in bathrooms, or in other high-moisture areas of a house. Preservative-treated plywood can withstand moisture better than interior plywood. Always check the local code(s) before beginning any construction project.

Interior plywood uses lower grades of veneer for the back and inner plies. Although the plies may be bonded with a water-resistant glue, waterproof glue is normally used. The lower-grade veneers reduce the bonding strength, however, which means that interior-rated panels are not suitable for exterior use.

High-Density Overlay (HDO) and Medium-Density Overlay (MDO) Plywood

High-density overlay (HDO) plywood panels have a hard, resin-impregnated fiber overlay heat-bonded to both surfaces. HDO panels are abrasion- and moisture-resistant, and can be used for concrete forms, cabinets, countertops, and similar high-wear applications. HDO plywood also resists damage from chemicals and solvents. HDO plywood panels are available in five common thicknesses: $\frac{3}{8}$", $\frac{1}{2}$", $\frac{5}{8}$", $\frac{3}{4}$", and 1".

Medium-density overlay (MDO) plywood panels are coated on one or both surfaces with a smooth, opaque overlay. MDO plywood accepts paint well and is very suitable for use as structural siding, exterior decorative panels, and **soffit** assemblies. MDO plywood panels are available in eight common thicknesses ranging from $1\frac{1}{32}$" to $2\frac{3}{32}$".

Both HDO and MDO panels are manufactured with waterproof adhesive and are suitable for exterior use. If MDO panels are to be used outdoors, however, the panels should be edge sealed with one or two coats of a good-quality exterior housepaint primer. An easy way to efficiently seal the edges is to stack panels and paint the edges of several panels at one time.

Face Veneers

Face veneers are the outermost plies of a plywood panel. Face veneer quality is indicated using the letters A, B, C, D, or N. Two letters separated by a hyphen are shown on a grade stamp to indicate the face veneer quality or grade. The quality of the front panel face is indicated by the first letter. The quality of the back panel face is indicated by the second letter.

Plywood that is designated for sheathing, subflooring, and concrete forms and panels used for special structural purposes are called *performance-rated plywood*. The veneers are either unsanded or lightly sanded. The plywood sheets carry slightly different grade markings. The face veneers are either C-C or C-D.

1.1.3 Engineered Wood Products

In the past, the primary source of structural beams, timbers, joists, and other weight-bearing lumber was old-growth trees. These trees, which need more than 200 years to mature, are tall and thick and can produce a large amount of high-quality, tight-grained lumber. Extensive logging of these trees to meet demand resulted in higher prices and conflict with forest conservation interests.

The development of wood laminating techniques by lumber producers has permitted the use of younger-growth trees in the production of structural building materials. These materials are given the general classification of engineered lumber products.

Engineered wood products fall into five categories: laminated veneer lumber (LVL), parallel strand lumber (PSL), laminated strand lumber (LSL), wood I-beams, and glue-laminated lumber or glulam (*Figure 4*).

Engineered wood products provide the following benefits:

- They can be made from younger, more abundant trees.
- They can increase the yield of a tree by 30 to 50 percent.
- They are stronger than the same size of structural lumber. Therefore, the same size piece of engineered lumber can bear more weight than that of solid lumber. Or, looked at another way, a smaller piece of engineered lumber can bear equal weight.
- Greater strength allows the engineered lumber to span a greater distance.
- A length of engineered wood is lighter than the same length of solid lumber. It is therefore easier to handle.

LVL is used for floor and roof beams and in scaffolding and concrete forms. It is also used to make the **header** over a window or door. No special cutting tools or fasteners are required.

PSL is used for beams, posts, and columns. It is manufactured in thicknesses up to 7". Columns can be up to 7" wide, and beams range up to 18" in width.

Soffit: Assembly that closes off the underside of the element of a building such as roof overhangs or beams.

Use of Engineered Wood Products

Engineered wood products are used in a wide array of applications that were once exclusively served by cut lumber. For example, PSL is used for columns, ridge beams, and headers. LVL is also used for form headers and beams. Wood I-beams are used to frame roofs as well as floors. An especially noteworthy application is the use of LSL studs, top plates, and soleplates in place of lumber to frame walls.

Header: A horizontal member that supports the load over an opening such as a door or window. Also known as a *lintel*.

Figure 4 Examples of engineered wood products.
Source: (A) Zoonar GmbH/Alamy Stock Photo; (B) Roy LANGSTAFF/Alamy Stock Photo; (C) Michael Doolittle/Alamy Stock Photo; (D) Justin Kase zsixz/Alamy Stock Photo; (E) REUTERS/Brendan McDermid; (F) Roy LANGSTAFF/Alamy Stock Photo

LSL is used for **millwork** such as doors and windows, and any other product that requires high-grade lumber. However, LSL will not support as great a load as a comparable size of PSL because PSL is made from stronger wood.

Wood I-beams consist of a web with flanges bonded to the top and bottom. This arrangement, which mimics the steel I-beam, provides exceptional strength. The web can be made of OSB or plywood. The flanges are grooved to fit over the web. A wood I-beam can be used as a floor joist, **rafter**, or header. Because of its strength, a wood I-beam can be used in greater spans than a comparable length of dimensional lumber.

Glulam is made from several lengths of solid lumber that have been glued together. It is popular in architectural applications where exposed beams are used (*Figure 5*). Because of its exceptional strength and flexibility, glulam can be used in areas subject to high winds or earthquakes.

Millwork: Various types of manufactured wood products such as doors, windows, and moldings.

Rafter: A sloping structural member of a roof frame to which sheathing is attached.

Figure 5 Glulam beam application.
Source: Michael Doolittle/Alamy Stock Photo

Cantilever: A beam, truss, or floor that extends beyond the last point of support.

Vaulted ceiling: A high, open ceiling that generally follows the roof pitch.

Glulam beams are available in widths from $2\frac{1}{2}$" to $8\frac{3}{4}$". Depths range from $5\frac{1}{20}$" to $28\frac{1}{2}$". They are available in lengths up to 40'. They are used for many purposes, including ridge beams; basement beams; headers of all types; and stair treads, supports, and stringers. They are also used in **cantilever** and **vaulted ceiling** applications.

Fire-Retardant Building Materials

Lumber and sheet materials are sometimes treated with fire-retardant chemicals. The lumber can either be coated with the chemical in a nonpressure process or impregnated with the chemical in a pressure-treating process. Fire-retardant chemicals react to extreme heat, releasing vapors that form a protective coating around the outside of the wood. This coating, known as *char*, delays ignition and inhibits the release of smoke and toxic fumes.

1.1.4 Building Boards

The ingenuity and technology that helped develop the plywood industry also assisted in the development of other materials in sheet form. The main ingredients for these products, known as *building boards*, are vegetable or mineral fibers. After mixing these ingredients with binder, the mixture becomes very soft.

At this point, the mixture passes through a press, which uses heat and pressure to produce the required thickness and density of the finished board.

Sawdust, wood chips, and wood scraps are the major waste materials at sawmills. These scrap materials are softened with heat and moisture, mixed with a binder and other ingredients, and then run through presses that produce the desired density and thickness.

The finished wood products that come off the presses are classified as hardboard, particleboard, or mineral fiberboard.

Hardboard

Hardboard is a manufactured building material, sometimes called *tempered board* or *pegboard*. Hardboards are water-resistant and extremely dense. The common thicknesses for hardboards are $\frac{3}{16}$", $\frac{1}{4}$", and $\frac{5}{16}$". The standard sheet size for hardboards is 4' × 8'.

These boards are susceptible to breaking at the edges if they are not properly supported. Holes must be predrilled for nailing; direct nailing into the material will cause it to fracture.

Three grades of hardboard are manufactured:

- *Standard* — Suitable only for interior use, such as cabinets.

- *Tempered* — The same as standard grade except that it is denser, stronger, and more brittle. Tempered hardboard is suitable for either interior or exterior uses such as siding, wall paneling, and other decorative purposes.

- *Service* — Not as dense, strong, or heavy as standard grade. It can be used for basically everything for which standard or tempered hardboard is used. Service grade hardboard is manufactured for items such as cabinets, parts of furniture, and perforated hardboard.

Particleboard

The main composition of this type of material is small particles or flakes of wood. Particleboard is pressed under heat into panels. The sheets range in size from $\frac{1}{4}$" to $1\frac{1}{2}$" in thickness. There are also thicknesses of 3" for special purposes. The standard size is 4' × 8'. Particleboard has no grain, is smoother than plywood, is more resilient, and is less likely to warp.

Underlayment: A material such as plywood or particleboard that is installed on top of a subfloor to provide a smooth surface for the finish flooring.

Some types of particleboard can be used for **underlayment** if permitted by the local building codes. If particleboard is used as underlayment, it is laid with the long dimension across the joists and the edges staggered. Particleboard can be nailed, although some types will crumble or crack when nailed close to the edges.

Mineral Fiberboard

The building boards just described are classified as vegetable fiberboards. Mineral fiberboards fall into the same category as vegetable fiberboards. The main difference is that they will not support combustion. Glass and **gypsum** rock are the most common minerals used in the manufacture of these fiberboards. Fibers of glass or gypsum powder are mixed with a binder and pressed or sandwiched between two layers of asphalt-impregnated paper, producing a rigid insulation board.

Some types of chemical foam mixed with glass fibers will also make a good, rigid insulation. However, this mineral insulation will crush and should not be used when it must support a heavy load.

Gypsum: A chalky type of rock that serves as the basic ingredient of plaster and gypsum board.

WARNING!

Whenever working with older materials that may be made with asbestos, contact your supervisor for the company's policies on safe handling of the material. State and federal regulations require specific procedures to follow prior to removing, cutting, or disturbing any suspect materials. Also, some materials emit a harmful dust when cut. Check the SDS before cutting. Asbestos can be found in structures built before 1978. It was used in ceiling tiles, siding, floor coverings, shingles, and pipe insulation.

1.2.0   Gypsum and Specialty Boards

Gypsum board, also known as *drywall*, is one of the most popular and economical methods of finishing the interior walls and ceilings of wood-framed and metal-framed buildings. Properly installed and finished, gypsum board can give a wall or ceiling made from many panels the appearance of being made from one continuous sheet (*Figure 6*).

Gypsum board is a generic name for products consisting of a noncombustible core. It is rated as limited combustible because of the paper. This product is made primarily of gypsum with a paper surfacing covering the face, back, and long edges. It is also called plaster board and Sheetrock®, a trade name of USG Corporation.

Gypsum board: A generic term for gypsum core panels covered with paper on both sides. It is commonly used to finish walls.

Figure 6 Typical gypsum board application.
Source: Levent Konuk/Shutterstock

1.2.1 Types of Gypsum Products

Many types of gypsum board are available for a variety of building needs (*Table 2*). Gypsum board panels are mainly used as the surface layer for interior walls and ceilings; as a base for ceramic, plastic, and metal tile; for exterior soffits; for elevator and other shaft enclosures; and to provide fire protection for architectural elements.

The following thicknesses of gypsum board are available in regular, Type X, and water-resistant boards.

- $\frac{1}{4}$" gypsum board — A lightweight, low-cost board used as a base in a multi-layer application for improving sound control and to cover existing walls and ceilings in remodeling. It is made in standard and high-flex (designed for curved areas, such as walls, arches, barrel ceilings, and stairways).
- $\frac{5}{16}$" gypsum board — A lightweight board developed for use in manufactured construction, primarily mobile homes.
- $\frac{3}{8}$" gypsum board — A lightweight board principally applied in a double-layer system over wood framing and as a face layer in repair and remodeling.
- $\frac{1}{2}$" gypsum board — Generally used for single-layer wall and ceiling construction in residential work and in double-layer systems to improve sound control and **fire rating**. The $\frac{1}{2}$" thickness can be used in fire-rated assemblies only when using **Type C**. Sag resistant board is also made in $\frac{1}{2}$" for ceiling applications.

Fire rating: A classification indicating in time (hours) the ability of a structure or component to withstand fire conditions.

Type C: A type of interior gypsum board that is rated as fire resistant and has more glass fibers than Type X, as well as unexpanded vermiculite components.

TABLE 2 Types and Uses of Gypsum Board

Type	Thickness	Standard Sizes[1]	Use
Regular, paper faced	$1/4$" $3/8$" $1/2$" $5/8$"	4' × 8' 4' × 8' 4' × 8' to 16' 4' × 8' to 16'	• Recovering old gypsum walls • Double-layer installation • Standard single-ply installation
Type X, fire-resistant	$5/8$"	4' width 8–12' lengths	• In garages, workshops, and kitchens • Around furnaces, fireplaces, and chimney walls • $5/8$" is used in fire-rated assemblies
Type C, fire-resistant	$1/2$", $5/8$"	4' width 8–12' lengths	• In garages, workshops, and kitchens • Around furnaces, fireplaces, and chimney walls • $5/8$" - is used in fire-rated assemblies
Water-resistant	$1/2$", $5/8$"	4' width 8–12' lengths	• For tile backing in areas not exposed to constant moisture • In kitchens, bathrooms, and utility rooms
Gypsum shaftliner	1"	2' width 8'–12' lengths	• Shaftliner partitions
Stretch gypsum board	$1/2$", $5/8$"	4.5' width 8–12' lengths	• Interior, 9' walls with extremely smooth finishes • In horizontal applications in conjunction with 9' ceilings
Mold-resistant	$1/2$", $5/8$"	4' widths 8–12' lengths	• In places where there is increased moisture • Can be installed anywhere in the interior or exterior
Abuse Resistant (AR)	$5/8$"	4' widths 8–12' lengths	• In environments where heavy impacts to the walls are expected • In areas where the surface needs to be protected, such as classrooms with lighter traffic, lobbies, boutique hotels, etc.
Impact Resistant (IR)	$5/8$"	4' widths 8–12' lengths	• In areas where the surface needs to be protected the most because it will be affected over time by crowds of people and moving equipment, such as mail rooms, hospital hallways, student housing, multi-family corridors, etc.
Glass mat gypsum board	$1/2$", $5/8$"	4' × 8', 3' x 5'	• Exterior sheathing • Interior walls subject to high water or moisture exposure

[1]For each of these boards, custom lengths are available from most manufacturers.

The Way It Was

Until the 1930s, walls were typically finished by installing thin, narrow strips of wood or metal known as *lath* between studs, and then coating the lath with wet plaster. Skilled plasterers could produce a very smooth wall finish, but the process was time-consuming and messy. In the early 1930s, paper-bound gypsum board was introduced and soon came into widespread use as a replacement for the tedious lath and plaster process.

Shaft wall: A system of nonbearing, fire-rated partitions used to enclose shafts or stairs that need fire and air resistance.

Type X: A type of interior gypsum board that is rated fire resistant and has a gypsum core with the inclusion of glass fiber strands.

- $5/8$" gypsum board — Used in quality single-layer and double-layer wall systems. The greater thickness provides additional fire resistance, higher rigidity, and better impact resistance. It is also used to separate occupied and unoccupied areas, such as a house from a garage or an office from a warehouse.

- 1" gypsum board — Used as a **shaft wall** liner and in semisolid or solid gypsum board partitions. It is also known as *coreboard*.

Standard gypsum boards are 4' wide and 8', 9', 10', or 14' long. The width is compatible with the standard framing of studs or joists spaced 16" or 24" on center.

Regular gypsum board is used as a surface layer on walls and ceilings. **Type X** gypsum board is available in $5/8$" thicknesses and has an improved fire resistance made possible by using special core additions. Type X gypsum board is used in most fire-rated assemblies.

Gypsum base for veneer plaster is used as a base for thin coats of hard, high-strength gypsum veneer plaster. This board is colored light blue.

Water-Resistant Gypsum Boards

Now that mold growth has been shown to be a severe environmental and health hazard, drywall manufacturers have improved the capability of gypsum board to resist it. Mold can grow on just about any surface, but there are gypsum boards that reduce absorption of moisture and slow down the growth of mold.

Figure 7 Water-resistant gypsum board.
Source: Kinek00/Alamy Stock Photo

Water-resistant gypsum board, also known as MM or MR (**mold or moisture-resistant gypsum board**) board, has a water-resistant gypsum core and water-repellent paper. The facing may have a light green, blue, or purple color depending on the manufacturer (*Figure 7*). Water-resistant gypsum board is not recommended for use in tubs and shower enclosures and other areas exposed to constant water.

Mold-resistant gypsum comes both as a 1" regular or Type X core board, and as 1/2" and 5/8" exterior boards. Some manufacturers chemically treat the papering on the gypsum to prevent mold growth. Rather than treating the paper, other manufacturers replace it with another mold-resistant material, such as glass mat facings. The sizes and installations of mold-resistant gypsum board are similar to those of regular board.

> **Mold or moisture-resistant gypsum board**: A type of water-resistant gypsum board that is marketed as either mold-resistant or moisture-resistant, or both.

WARNING!

When handling mold-resistant gypsum board, be sure to wear proper PPE, including hand, face, and respiratory protection. The chemical treatments and cellulose fibers can get on your skin or be inhaled, which may cause irritation or health problems.

Tile Backer Boards

A **tile backer board** is a special type of board that is replacing water-resistant gypsum board as a backing for tile in damp areas such as baths and shower stalls. There are a few types of tile backer board.

Glass mat gypsum board has a moisture-resistant fiberglass reinforced gypsum core and a coated fiberglass mat face on both sides that acts as a water barrier. It is more resistant to mold and moisture than the regular, paper-faced gypsum board. It can be used both as exterior sheathing and for interior walls in areas that are exposed to more liquid and moisture. It can be used to underlay tiles and tends to be easier to cut and handle than cement board.

Another type of tile backer board is known as **cement board** (*Figure 8*). It is made from a slurry of portland cement mixed with reinforcing fibers. It is colored light blue for easy recognition. These backer boards, in addition to their use as a tile backer, can be used as a floor underlayment, countertop base, heat shield for stoves, and as a base for exterior finishes such as **stucco** and brick veneer. They are available in 32", 36", and 48" widths and 5' and 8' lengths. Common thicknesses are 1/4" and 1/2".

> **Tile backer board**: A board that is used to underlay ceramic tiles in wet areas of a building.
>
> **Glass mat gypsum board**: A type of gypsum board with a moisture-resistant core and face that allows it to act as a water barrier in interior and exterior applications.
>
> **Cement board**: A type of durable tile backer board made of cement and reinforcing fibers.
>
> **Stucco**: A type of plaster used to coat exterior walls.

Gypsum shaftliner: A type of heavy-duty fire- and moisture-resistant gypsum core panel that is used in shafts, stairwells, and chutes.

Gypsum Shaftliner

Gypsum shaftliner panels are specialized panels with noncombustible, Type X, moisture-resistant gypsum cores inside heavy-duty paper facers or fiberglass mats. They are available as a 1"-thick solid core board and provide fire protection for interior shafts, stairs, and chutes. Gypsum shaftliner panels are used in shaft walls and shaftliner partitions with additional layers of gypsum board applied to the core board to complete the wall assembly. They are available in a width of 24" and with a variety of edges, the most common being square and beveled.

54" Board

Stretch board panels come 6 inches wider than the standard 4'. These panels are designed for backing the thinner wallpapers and no-texture finishes done in newer construction. Two standard-width boards come up one foot short when hanging drywall on nine-foot walls, but 54" boards are wide enough to reduce the number of drywall joints and make a smoother surface.

Gypsum Sheathing

Gypsum sheathing is used as a protective, fire-resistant membrane under exterior wall surfacing materials such as wood siding, masonry veneer, stucco, and shingles. It also provides protection against the passage of water and wind and adds structural rigidity to the framing system. The noncombustible core is surfaced with firmly bonded, water-repellent paper. In addition, a water-repellent material may be incorporated in the core. It is available in 2' and 4' widths, and $\frac{1}{2}$" to $\frac{5}{8}$" thicknesses. The $\frac{5}{8}$"-thick boards are available with Type X cores.

Abuse Resistant (AR) gypsum panels: Gypsum panels designed to withstand minor abrasions and impacts against a wall.

Impact Resistant (IR) gypsum panels: Gypsum panels designed to withstand stronger or more frequent impacts against a wall than AR panels.

Acoustical gypsum board: Gypsum board that is designed to provide enhanced sound control and soundproofing.

AR and IR Gypsum Panels

Abuse Resistant (AR) gypsum panels are designed to withstand minor abrasions and impacts against the wall, and to prevent indentations and damage from objects. They have strong face paper and backing, and can be made with a gypsum-cellulose fiber core. AR gypsum panels are more rigid and have higher structural integrity.

   Impact Resistant (IR) gypsum panels are similar to AR panels, but are designed to withstand stronger or more frequent impact against the wall. For instance, you could install an AR panel in an airport hallway that gets light traffic, but you should install an IR panel in an airport hallway that gets heavy traffic.

Acoustical Gypsum Board

Acoustical gypsum board is a type of gypsum board that is made specifically for areas that need enhanced sound attenuation or sound control. It is made with a

Figure 8 Cement board behind bathroom tiles.
Source: Jeffrey Isaac Greenberg 10+/Alamy Stock Photo

Special-Use Gypsum Board

Regular $\frac{1}{2}$" and $\frac{5}{8}$" gypsum board are the most common types. Several types of gypsum board are designed for special applications, however, including the following:

- Type X gypsum board provides improved fire ratings because its core material is mixed with fire-retardant additives. Type X is often used on walls that separate occupancies. Examples are walls and ceilings between apartments or a wall separating a garage from the living area of a house. Use of Type X is normally specified by local building codes for protection of occupants.
- Type C is an enhanced version of Type X. It is made with more noncombustible glass fibers than Type X as well as unexpanded vermiculite components.
- Flexible $\frac{1}{4}$" drywall panels have a heavy paper face and are designed to bend around curved surfaces.
- Special high-strength drywall panels are made for ceiling applications. The core of these panels is specially treated to resist sagging.
- A weather-resistant drywall panel is available for installation on soffits, porch ceilings, and carport ceilings.

   Because they have a water-resistant core covered on both sides with water repellant paper, Type C and Type X gypsum sheathing panels are used in cases where the required fire rating of exterior walls exceeds that available with OSB, plywood, or other types of sheathing. Gypsum sheathing panels are widely used in commercial construction.

viscoelastic gel layer between two thin layers of gypsum. These boards are often used to meet certain enhanced building requirements for sound in the United States, such as STC and ASTC.

1.2.2 Magnesium Oxide (MgO) Boards

Magnesium oxide (MgO) boards, or MgO boards, are another type of building board that has been developed in recent years. They provide a greener alternative to gypsum panels and other construction boards in a wall assembly (*Figure 9*). They can be applied in a variety of ways in a building and used as sheathing, trim, exterior siding, tile backer, wall panel, ceiling board, or firewall. They are installed in the same way as gypsum board.

One downside to using MgO boards is that they tend to have a larger carbon footprint to manufacture because most of the raw materials must be transported from China. The other main downside is the cost, which is higher than that of gypsum board.

Magnesium oxide (MgO) boards: A type of construction board that can be installed for a variety of uses, such as interior wall or ceiling panels or exterior sheathing, siding, or trim.

Figure 9 MgO board.
Source: ILIA NEZNAEV/Shutterstock

1.3.0 Masonry Materials

For the purposes of this module, the term *masonry* includes construction using stone, brick, concrete block, and poured concrete. These materials are used extensively in residential and commercial construction, and it is important for drywall mechanics to know how they work as part of the wall assembly. Special tools and fasteners are used with these materials.

1.3.1 Concrete

Concrete is a mixture of four basic materials: portland cement, fine aggregates, coarse aggregates, and water. Various types of concrete can be obtained by varying the basic materials and/or by adding other materials to the mix. When first mixed, concrete is in a semi-liquid state and is referred to as **plastic concrete**. When the concrete hardens, but has not yet gained structural strength, it is called **green concrete**. After the concrete has hardened and gained its structural strength, it is called *cured concrete*.

In residential construction, concrete may be used in foundation walls and footings, basement floors, or as the foundation slab if the house has no basement. In commercial construction, the entire structure, including floors, walls,

Plastic concrete: Concrete in a liquid or semi-liquid workable state.

Green concrete: Concrete that has hardened but has not yet gained its full structural strength.

and support columns, may be made of concrete. Walls can be anywhere from a few inches to several feet thick.

> **WARNING!**
>
> Those working with or around cement should be aware that it is harmful. Dry and wet cement can cause blood poisoning and chemical burns. Make sure to wear the appropriate personal protective equipment when working with dry or wet concrete. If wet concrete enters waterproof boots from the top, remove the boots and rinse your legs, feet, boots, and clothing with clear water as soon as possible. Repeated contact with cement or wet concrete can cause an allergic reaction in certain individuals.

1.3.2 Concrete Masonry Units (CMUs)

Commonly known as concrete block, concrete masonry units (CMUs) are one of the most common building materials in both residential and commercial construction. They are made from a mixture of portland cement, aggregates such as sand and gravel, and water.

Hollow concrete block is used in all kinds of residential and commercial applications. Residential basement walls are usually made of concrete block, and it is often used as a base for finish materials such as brick and stucco.

1.3.3 Brick

Brick is commonly used as a veneer for residential and commercial buildings. Brick is made from pulverized clay that is mixed with water and then molded into various shapes, primarily rectangular. Once the brick hardens and dries, it is fired in a furnace to provide the necessary hardness.

Like cement block, bricks are bonded together with mortar. Brick is typically laid against a supporting structure such as a concrete block wall or a frame wall sheathed with plywood (*Figure 10*). It might connect with the continuous insulation on the outside of the exterior wall assembly. An air space is maintained between the two walls to allow moisture to escape. A weep hole is provided to drain condensation that develops in the air space. The main difference between conventional frame construction and brick facing is that the foundation wall is extended to provide support for the brick.

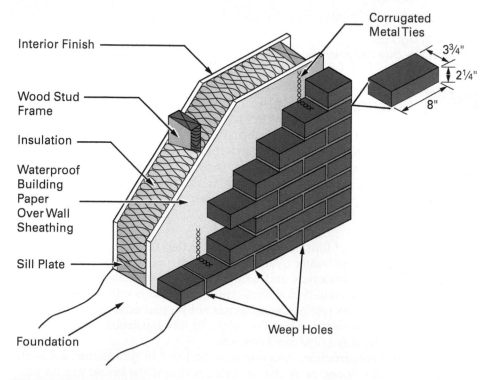

Figure 10 Brick veneer wall.

Brick veneer is commonly used in midrise construction with cold-formed steel framing. This requires special detailing, such as brick ties to allow for differential movement of structures.

1.3.4 Stone

Like brick, stone is used primarily as a facade over block or frame walls. Stone used for this purpose can be as much as 6" thick. However, in renovating very old homes, you may find stone foundations and walls a foot or more thick, and they are very difficult to drill through.

1.4.0   Metal Materials

Metals have a variety of applications, especially in commercial construction.

1.4.1 Steel

Cold-Formed Steel

Light gauge steel studs are used in framing walls, floors, and roofs. Light gauge steel is generally made with **cold-formed steel (CFS)**, which is made of sheet steel. The steel is rolled into C- or Z-section shapes in a process that does not require heat. Because CFS studs can hold more weight than wood studs, they can be placed further apart, resulting in fewer studs. This may affect the amount and sizing of the drywall, insulation, and other materials needed.

CFS studs usually come with pre-punched holes (punchouts) for mechanical, plumbing, and electric installation. They can be fastened using steel screws.

Cold-formed steel (CFS): A type of steel made of sheet steel in a process that doesn't involve heat.

Structural Steel

In many commercial buildings, the horizontal and vertical support member is a heavy gauge structural steel **girder** and beam (*Figure 11*). Structural steel is hot rolled in a very different manufacturing process to light gauge steel. It is stronger and heavier and needs to be fastened with welding, riveting, or bolting.

Girder: The main steel or wood supporting beam for a structure.

Figure 11 Structural steel frame.
Source: since1985/123RF

1.4.2 Other Metal Materials

Screws and nails are some of the most important tools used in construction. Drywall mechanics may need to use one or both, based on the project and need.

Metal sheet material is common in walls and roofs of commercial buildings. Aluminum or tin foil might also be attached to insulation to create reflective insulation.

Corrugated: Material formed with parallel ridges or grooves.

Corrugated steel decking is used as a base for poured concrete floors in multistory commercial buildings. Steel reinforcing bars and mesh are used to strengthen poured concrete in all applications.

1.0.0 Section Review

1. When is drywall normally installed?
   a. After the building has been dried in
   b. After the walls have been painted
   c. Before insulation has been installed
   d. Before the electrical rough in

2. _____ gypsum boards are used for fire-resistant-rated assemblies.
   a. Water-resistant and Type C
   b. Type X and abuse-resistant
   c. Type X and Type C
   d. Water-resistant and Type X

3. Where might concrete be used in a residential building?
   a. Roof
   b. Foundation slab
   c. Interior walls
   d. Ceiling

4. Which material is used as a horizontal or vertical support member in commercial buildings?
   a. Brick veneer
   b. Corrugated steel decking
   c. Gypsum board
   d. Structural steel

2.0.0 Walls

Performance Tasks

This is a knowledge-based module. There are no Performance Tasks.

Objective

Explain wall construction processes.
   a. Describe wood frame wall construction.
   b. Describe steel frame wall construction.
   c. Describe differences in commercial and residential walls.

Drywall mechanics must know the basic components of wall construction and framing. Insulation and gypsum board are installed around the framing.

Wood frame construction (*Figure 12*) has been in common use since the 1800s. Frame construction begins by building a foundation, which usually consists of a poured concrete **footing**. In cold climates, the footing must be built below the frost line to prevent it from cracking. If there is no basement, a short foundation wall of poured concrete is set onto poured concrete footings.

Footing: The foundation for a column or the enlargement placed at the bottom of a foundation wall to distribute the weight of the structure.

Steel frame construction (*Figure 13*) has also been in use since the 1800s. The first documented use of cold-formed steel (CFS) in building construction was in a Virginia hospital in 1925. Steel framing gained lasting popularity in commercial construction in the United States in the 1950s and 1960s. Steel framing is code mandated in nonresidential construction because it is noncombustible. Because you can't use nails with steel framing, it was the invention of the drywall screw that allowed gypsum panels to be efficiently fastened to the stud.

Figure 12 Example of a wood frame building.
Source: Andy Dean Photography/Shutterstock

Figure 13 Example of a steel frame building.
Source: zhengzaishuru/Shutterstock

Today, 30 to 35 percent of all nonresidential buildings in the United States are made with CFS. Drywall mechanics will encounter both wood and CFS framing in their work and must know how to work with both. Some structures are also built with hybrid framing, which is a combination of both wood and CFS framing.

2.1.0 Wood Wall Framing

Wood wall framing is generally done with 2 × 4 studs spaced 16" OC. In many cases, 24" spacing is used on interior walls. Some codes permit 24" spacing on exterior walls for one-story buildings. If 24" spacing is used in a two-story building, the lower floor must be framed with 2 × 6 lumber.

Figure 14 identifies the structural members of a wood frame wall. Each of the members shown on the illustration is described as follows:

- **Blocking** *(spacer)* — A wood block that is used as a filler piece and support between framing members. Blocking also provides a surface for attaching equipment and other items.
- **Cripple stud** — A short framing stud that fills the space between a header and a **top plate** or between the sill and the soleplate.
- **Double top plate** — A plate made of two members to provide better stiffening of a wall. It is also used for connecting splices, corners, and partitions that are at right angles (perpendicular) to the wall.
- *Header (lintel)* — A horizontal structural member that supports the load over an opening such as a door or window.
- **King stud** — The full-length stud next to the **trimmer stud** in a wall opening.
- *Partition* — A wall that subdivides space within a building. A bearing partition or wall is one that supports the floors and roof directly above in addition to its own weight.
- *Rough opening* — An opening in the framing formed by framing members, usually for a window or a door.
- *Rough sill* — The lower framing member attached to the top of the lower cripple studs to form the base of a rough opening for a window.
- *Soleplate* — The lowest horizontal member of a wall or partition to which the studs are nailed. It rests on the rough floor.
- *Common stud* — The main vertical framing member in a wall or partition.

NOTE

Codes may require variations in structure for seismic and other hazards.

Blocking: A wood block used as a filler piece and support member between framing members.

Cripple stud: In wall framing, a short framing stud that fills the space between the header and the top plate or between the sill and the soleplate.

Top plate: The upper horizontal member of a wall or partition frame.

Double top plate: A length of lumber laid horizontally over the top plate of a wall to add strength to the wall.

King stud: The full-length stud next to the trimmer stud in a wall opening.

Trimmer stud: The vertical framing member that forms the sides of a rough opening for a door or window. It provides stiffening for the frame and supports the weight of the header.

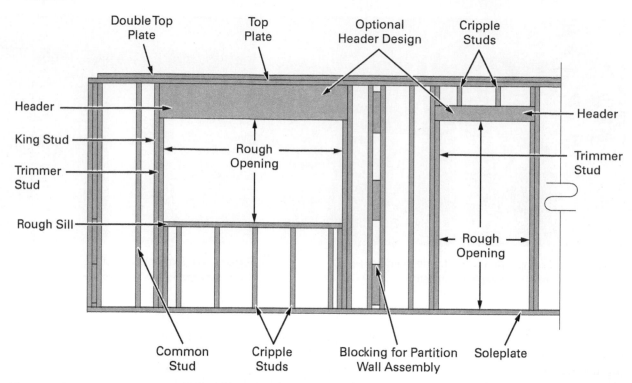

Figure 14 Wood frame wall and partition framing members.

Truss: An engineered assembly made of wood or metal that is used in place of individual structural members such as the joists and rafters used to support floors and roofs.

- *Top plate* — The upper horizontal framing member of a wall used to carry the roof **truss** or rafter.
- *Trimmer stud* — The vertical framing member that forms the sides of rough openings for doors and windows. It provides stiffening for the frame and supports the weight of the header.

2.1.1 Corners

A wall must have solid corners that can take the weight of the structure. In addition to contributing to the strength of the structure, corners must provide a good nailing surface for sheathing and interior finish materials. Building contractors generally select the straightest, least defective studs for corner framing.

There are many methods for constructing corners (*Figure 15*). Some builders will construct the corner in place, then plumb and brace it before raising the wall frames. This approach makes it easier to plumb and brace the frame, but it prevents installation of the sheathing before the frame is erected. If the corners are included in the frame, then a portion of the corner is included with each of the mating frame sections.

2.1.2 Partition Intersections

Interior partitions must be securely fastened to outside walls. A solid nailing surface is required where the partition intersects the exterior frame. There are several methods used to construct framing for partition Ts (*Figure 16*).

2.1.3 Window and Door Openings

When wall framing is interrupted by an opening such as a window or door, a method is needed to distribute the weight of the structure around the opening. This is done by using a header (*Figure 17*). The header is placed so that it rests on the trimmer studs, which transfer the weight to the soleplate or subfloor and then to the foundation.

Headers are made of solid or built-up lumber. Laminated lumber and beams have become popular as header material, especially where the load is heavy.

Other types of headers used for heavy loads are wood or steel I-beams and box beams. The latter are made of plywood webs connected by lumber flanges in a box configuration.

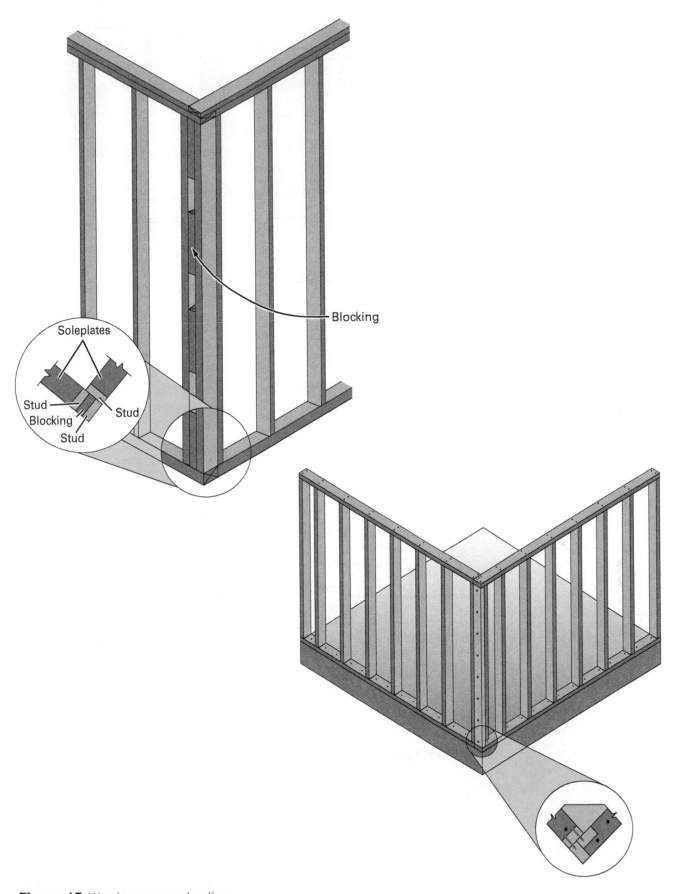

Figure 15 Wood corner construction.

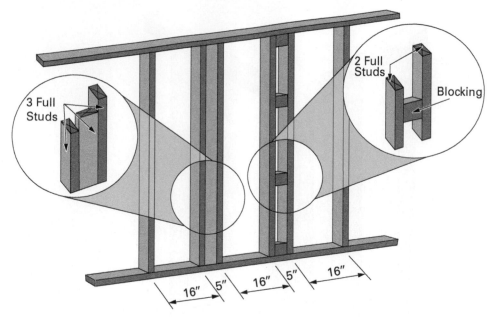

Figure 16 Constructing nailing surfaces for partitions.

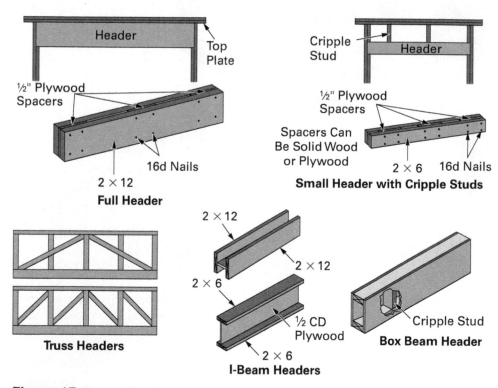

Figure 17 Types of headers.

Figure 18 shows cross sections of typical wood-framed walls.

2.1.4 Bracing

Bracing is important in the construction of exterior walls. Many local building codes require bracing if certain types of sheathing are used. In areas where high winds or earthquakes are a hazard, additional wind bracing may be required even when $\frac{1}{2}$" plywood is used as the sheathing. Fastener spacing will vary by design but is essential for the bracing to function.

Several methods of bracing have been used since the early days of construction. One method is to cut a notch (let-in) for a 1 × 4 or 1 × 6 at a 45-degree angle on each corner of the exterior walls. Another method is to cut 2 × 4 braces at a

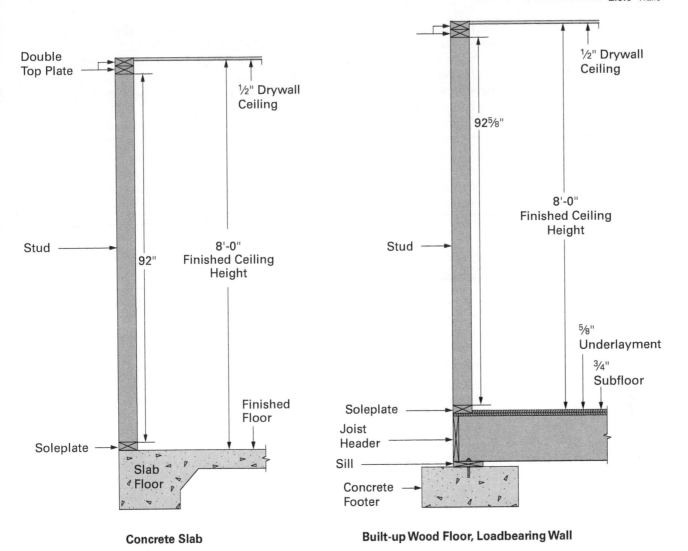

Figure 18 Cross sections of wood-framed walls.

45-degree angle for each corner. Still another type of bracing (used where permitted by the local code), is metal strap bracing (*Figure 19*). This product is made of galvanized steel.

Metal strap bracing is easier to use than let-in wood bracing. Instead of notching out the studs for a 1 × 4 or 2 × 4, a circular saw is used to make a diagonal groove in the studs, top plate, and soleplate for the rib of the bracing strap. The strap is then nailed to the wood framing.

With the introduction of plywood, some areas of the country have done away with corner bracing. However, along with plywood came different types of sheathing that are byproducts of the wood industry and do not have the strength to withstand wind pressures. When these are used, permanent bracing is needed. Building codes in some areas will allow a sheet of ½" plywood to be used on each corner of the structure in lieu of diagonal bracing when the balance of the sheathing is fiberboard. In other areas, the use of bracing is required regardless of the type of sheathing used. Always check local codes.

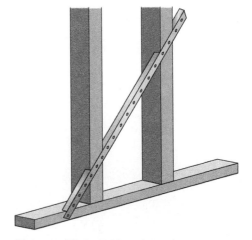

Figure 19 Metal bracing.

2.1.5 Sheathing

Sheathing is the material used to close in the walls. Plywood and non-veneer panels are generally used for sheathing.

When plywood is used, the panels will range from ⁵⁄₁₆" to ¾" thick. A minimum thickness of ³⁄₈" is recommended when siding is to be applied. The higher end of the range is recommended when the sheathing acts as the exterior finish surface. The panels may be placed with the grain running horizontally or vertically. If they are placed with the grain running horizontally, local building codes may require that blocking be used along the top edges.

Nails must be used when attaching sheathing to framing, as specified by *IBC*® and *IRC*®. Typical nailing requirements call for 6d (6 penny) nails for panels $\frac{1}{2}$" thick or less and 8d nails for thicker panels. Fastener spacing is critical and will vary with location, local code, and drawings.

Other materials that are sometimes used as sheathing are fiberboard (insulation board), rigid foam sheathing, and exterior-rated gypsum board. A major disadvantage of these materials is that siding cannot be nailed to them. It must either be nailed to the studs or special fasteners must be used.

When material other than rated panels is used as sheathing, rated plywood panels may be installed vertically at the corners to eliminate the need for corner bracing.

2.2.0   Steel Wall Framing

Some aspects of steel wall construction are very different from wood wall construction. But, despite differences in materials, approach, and measurements, the overall structure of steel and wood-framed walls is still similar.

The advantages of CFS framing include incombustibility, uniformity of dimension, lightness of weight, freedom from rot and moisture, termite resistance, and ease of construction. The components of CFS frame systems are manufactured to fit together easily. There are a variety of CFS framing systems, both load-bearing and nonbearing types.

2.2.1 CFS Framing Materials

CFS framing components include CFS studs and, in some cases, metal joists and roof trusses (*Figure 20*). The vertical and horizontal framing members serve as structural load-carrying components for a variety of low- and high-rise structures. CFS stud framing is compatible with all types of surfacing materials that can be applied with screws.

Figure 20 Metal trusses.
Source: Beekeepx/Shutterstock

Track: A length of steel that goes on the top and bottom of the wall or ceiling to receive steel stud framing members.

A steel wall is built using a ceiling and floor **track** or runner (*Figure 21*). The floor track is fastened to the ground, and studs are cut to fit in the room and attached to the tracks with screws. They are generally spaced 16" or 24" apart. The type of steel used for framing is light gauge and is not designed to be load-bearing.

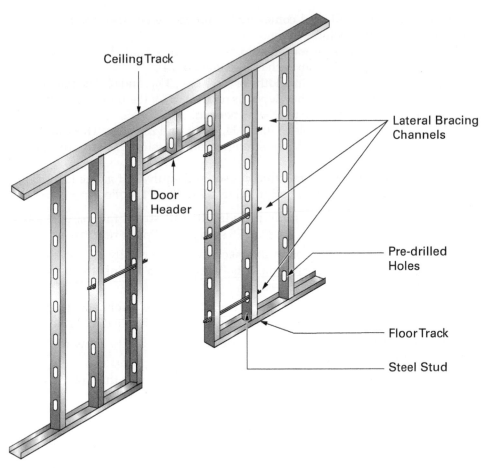

Figure 21 Steel frame wall and partition framing members.

When CFS studs are used for drywall framing systems, the channel stock comes in two grades. The first grade is a **non-structural stud** or standard **drywall stud** (*Figure 22*). Standard studs come in widths of $1\frac{5}{8}$", $2\frac{1}{2}$", $3\frac{5}{8}$", and 6". The flange is $1\frac{1}{4}$" with a $\frac{3}{16}$" return lip. Standard lengths are 10', 12', and 14'; other lengths are available by special order. A CFS stud taking the place of a standard wood 2×4 stud would be $3\frac{5}{8}$". A CFS stud taking the place of a wood 2×6 would be 6".

Non-structural stud: A nonbearing stud for walls with gypsum sheathing.

Drywall stud: A type of CFS stud that is between 18 and 33 mils and is meant for nonbearing walls.

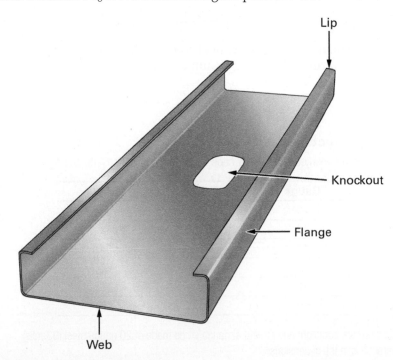

Figure 22 Standard CFS stud stock.

NOTE

The term *gauge* has traditionally been used to state the thickness of steel stud material. In recent years, however, the industry has converted to using mils, which represents the thickness of the steel in millimeters.

TABLE 3 Minimum Base Steel Thickness of Cold-Formed Steel Members

Designation Thickness (Mils)	Reference Gauge Number
18	25
27	22
30	20–Drywall
33	20–Structural
43	18
54	16
68	14
97	12
118	10

Source: Adapted from the Steel Framing Industry Association (SFIA) specifications: https://sfia.memberclicks.net/assets/TechFiles/SFIA%20Tech%20Spec%202015%20updated%207.12.17%20v.2.pdf

Structural stud: Extra heavy CFS or wood stud that is used in the exterior or structural frame of a building, is load-bearing, and is used to resist environmental loads.

Slap stud: The last stud of an intersection wall, typically a T-intersection or inside corner at an intersection.

Deflection track: A track used for CFS studs typically in exterior walls to deflect the roof or floor load above without transferring axial load to the studs.

Furring: Strips of wood or metal applied to a wall or other surface to make it level, form an air space, and/or provide a fastening surface for finish covering.

Z-furring channel: A Z-shaped steel channel that is used to furr out walls and to create a uniform surface for gypsum board.

The standard drywall stud comes in 20 and 25 gauge steel (the higher the gauge, the lighter the metal).

The second grade is a **structural stud**. Studs of this grade have cutouts and utility knockout holes 12" from each end and every 24".

Structural studs also come in widths of $1\frac{5}{8}$", $2\frac{1}{2}$", $3\frac{5}{8}$", and 6". The flanges are $1\frac{3}{8}$" to $3\frac{1}{2}$" with a return lip that depends on the size of the flange. This type of stud can be ordered in any length that is needed.

Structural studs are available in 10, 12, 14, 16, 18, and 20 gauge. These are typically used on exterior work. These strengths are classified as structural steel studs.

Table 3 lists common base steel thicknesses and equivalent gauge identifiers. Some commonly used terms in CFS framing are defined as follows:

- *Stud* — For CFS framing, studs include all wall studs, rafters, and joists because they share the same shape. CFS studs have lips or returns.
- **Slap stud** — The last stud of an intersection wall, typically a T-intersection or inside corner at an intersection. It is "slapped" up against the bypassing drywall. It is also known as a *floater*.
- *Flange* — The edge of the stud that connects to the web on either side and the area where the drywall or other enclosing material is applied to the stud.
- *Knockout* — A hole that has been pre-cut into a CFS stud to allow for utility lines. Also known as a *punchout*.
- *Lip* — An extension of the flange that curls slightly over the web on either side and the part of the stud that connects with the screw.
- *Web* — The back and widest part of the stud.
- *Track or runner* — A length of steel that runs on the floor or ceiling to which the other steel stud framing members will be attached.
- **Deflection track** — A track used for CFS studs in exterior walls to deflect the roof or floor load above without transferring axial load to the studs.
- **Furring** channel — A hat-shaped length of steel (also known as a *hat channel*) that is attached to studs and used to raise or level walls or ceilings. It can also be used to create a gap between studs and drywall to prevent moisture buildup.
- *L-headers* — CFS members that can be used as headers. They can be doubled to create the strength needed for headers.
- *U-channel* — A CFS material that can be used for blocking, bridging, and cabinet backing.
- **Z-furring channel** — A Z-shaped steel channel that is used to furr walls and to create a uniform surface for gypsum board.

Identifying Structural Studs

Structural studs are marked with a color code for easy identification. The coding is as follows:

Gauge	Color	Mils
20	White	33
18	Yellow	43
16	Green	54
14	Orange	68
12	Red	97
10	Blue	118

Note that there are both light and structural gauge studs made of 20 gauge steel (33 mils). The difference is in the dimensions.

2.2.2 The CFS Framing Process

Although different materials are used, the general approach to framing with CFS studs is the same as that used for wooden studs.

Like wood framing, CFS framing is installed 12", 16", or 24" OC, openings are framed with headers and cripples, and special framing is needed for corners and partition Ts.

Depending on the load, reinforcement may be needed when framing openings. Bracing of walls to keep them square and plumb is also required. The illustrations in this section show examples of common framing techniques. *Table 4* shows the framing spacing for various gypsum board applications.

The erection of CFS studs typically starts by laying metal tracks in position on the floor and ceiling and securing them (*Figure 23*). If the tracks are being applied to concrete (*Figure 24*), a low-velocity powder-actuated fastener is generally used. If the tracks are being applied to wood joists, such as in a residence, screws can be driven with a screw gun. Always check the shop drawings for such details.

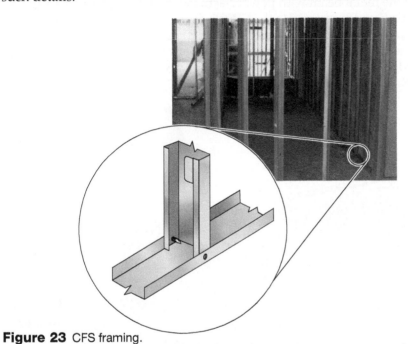

Figure 23 CFS framing.

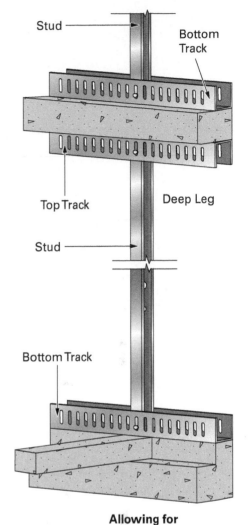

Allowing for Vertical Movement

Figure 24 CFS studs with concrete floors and ceiling.

TABLE 4 Maximum Framing Spacing

	Single-Ply Gypsum Board Thicknesses		Application to Framing		Maximum OC Spacing of Framing
Ceilings	3/8"		Perpendicular or Parallel		16"
	1/2"		Parallel		16"
	1/2"		Perpendicular		24"
	5/8"		Perpendicular		24"
Sidewalls	1/2" or 5/8"		Perpendicular or Parallel		24"
Fasteners Only–No Adhesive Between Plies					
	Multi-Ply Gypsum Board Thicknesses		Application to Framing		Maximum OC Spacing of Framing
	Base	Face	Base	Face	
Ceilings	3/8"	3/8"	Perpendicular	Perpendicular*	16"
	3/8"	3/8"	Parallel	Perpendicular	24"
Sidewalls	1/2" or 5/8"	1/2" or 5/8"	Perpendicular	Perpendicular	24"
	1/2" or 5/8"	1/2" or 5/8"	Perpendicular	Perpendicular	24"

*Must use adhesive

┌───┐
│ **WARNING!** │
└───┘

The use of a powder-actuated fastener requires special training and certification. The use of these tools may be prohibited by local codes because of safety and seismic concerns.

As shown in *Figure 24*, the stud is attached to slots in the top track. This allows for vertical movement. If the tracks or studs need to be cut to size, they can be cut with aviation snips or another type of sheet metal cutting tool.

Once the tracks are in place, the studs and openings are laid out in the same way as a wood frame wall. The studs may be secured to the tracks with screws, or they may be welded. In some cases, the entire wall will be laid out on the floor, then raised and secured. When heavy gauge walls are used, they may be assembled and welded in a shop and brought to the site.

When CFS studs are used to frame around steel beams, the CFS studs are secured to the metal beam with powder-actuated fasteners (if allowed) and the support members are screwed to the CFS studs (*Figure 25*). Note the wire hanger for the suspended ceiling at the right of the picture.

When the CFS studs are installed against metal channels or flanges, the studs are secured to the channel with scrap pieces of CFS studs (*Figure 26*). Remember to stay within the spacing and go no more than 16" or 24" OC.

Figure 25 Plate attached to a beam.

Figure 26 Studs secured to a channel.

2.2.3 Corners

Corners in CFS framing are planned when the tracks are laid out. The tracks overlap, and there are multiple studs for each track—two facing out to each side of track and an extra one behind them for reinforcement. A similar process is used for other types of intersections. *Figure 27* shows a CFS corner with a piece of gypsum board over one section of framing.

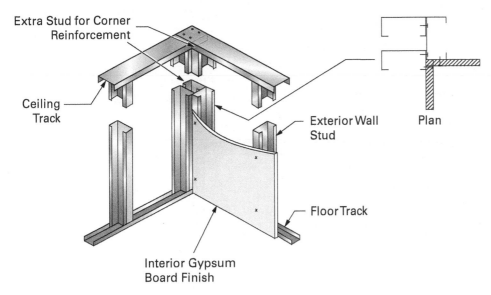

Figure 27 CFS corner construction.

2.2.4 Partition Intersections

There are some differences between installing metal nonbearing partitions and wooden nonbearing partitions. When building with wood, all partitions must be nailed together. With CFS studs, this is not required.

As shown in *Figure 28,* the partitions are held back from the other partitions so that the drywall will slide past.

Figure 28 Partition held back to let drywall slide by.

2.2.5 Window and Door Openings

Steel framing installations often require custom door frames provided by the manufacturer. These are either welded frames (pre-welded together by the manufacturer) or knock down (KD) frames that come in pieces and can be attached on site. Several steel studs need to be installed in the wall to accommodate the door—short jamb studs that extend vertically down from the ceiling track and a header stud that acts as a horizontal track for the jamb studs. The door frame will be anchored to this header stud and to the main steel studs on either side (*Figure 29*).

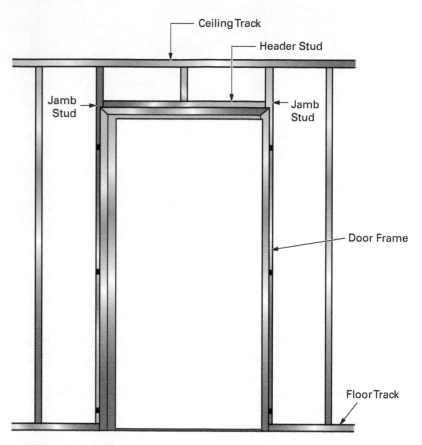

Figure 29 Door frame in a steel-framed wall.

Box header: A type of header that is made by combining a structural panel with the framing.

An alternative to these is a **box header**. This type of header can be made by installing a piece of track on top and bottom with multiple studs in between, laid on their edge to create a stronger header.

Windows will have similar openings that need to be built into the steel frame to allow for the size of the window's rough opening dimensions.

2.2.6 Bracing

While there are different bracing requirements depending on whether studs are load-bearing, all steel framing members should be braced in some way. Without bracing, the frames may twist, buckle, or fail. It is critical to prevent this from happening, particularly as the height of steel-framed buildings gets higher and higher.

Drywall mechanics must know about bracing because they may need to brace walls with gypsum boards or sheathing. Manufacturing tables and design notes should always be consulted when designing and installing bracing.

Sheathing braced design: A type of CFS stud bracing where the studs are braced by a sheathing material attached to one or both sides of the stud.

Steel braced design: A type of CFS stud bracing where the studs are braced by steel, whether by running the steel through the punchouts, strapping it across the outside of the studs, or adding blocks between each stud.

There are two basic methods of bracing CFS studs, as defined by the American Iron and Steel Institute (AISI): **sheathing braced design** and **steel braced design**. In sheathing braced design, the studs are braced by a sheathing material attached to one or both sides of the stud. This is generally the more cost-effective method of bracing because sheathing will be placed outside the studs anyway. However, it is not as stable and secure as steel bracing. Sheathing braced design may use the maximum axial nominal loads specified by the AISI in *Table 5*.

TABLE 5 Maximum Axial Nominal Load by Gypsum Board Sheathing–Wall Stud Connection Capacity

Gypsum Board Sheathing	Screw Size	Maximum Nominal Stud Axial Load
$1/2''$	6	5.8 kips (25.8 kN)
$1/2''$	8	6.7 kips (29.8 kN)
$5/8''$	6	6.8 kips (30.2 kN)
$5/8''$	8	7.8 kips (34.7 kN)

Source: AISI S211 Table B1-1 Maximum Axial Nominal Load Limited by Gypsum Sheathing-to-Wall Stud Connection Capacity

In steel braced design, the studs are braced by steel, whether by running the steel through the punchouts, strapping it across the outside of the studs, or adding blocks between each stud.

Load-Bearing Walls

Load-bearing walls are walls in the exterior or structural frame of a building that need to support environmental loads as well as the load (or weight) of the building.

Different forms of steel braced design are used to support CFS stud walls. Lateral bracing is always recommended as the minimum support for CFS stud walls. Lateral bracing is recommended even if the wall will later be covered in gypsum panels to ensure the framing is braced under the construction load that will be applied to it for weeks before gypsum boards are installed. Also, there is always the risk that gypsum board will be damaged by fire, water, or other environmental factors. For axially loaded walls, "all-steel" bracing is required because structural sheathing and gypsum board is not considered stable enough.

Lateral bracing can involve bridging, or running steel channels horizontally through the punchouts in the studs (*Figure 30*) and welding or screwing them with angle clips. The punchouts must align horizontally for this method to work. The channels are then mechanically attached to the studs with sheet steel angles. This is a type of tension-compression system, meaning it can resist stud buckling in tension or compression.

Lateral bracing can also be achieved through metal strap bracing (*Figure 31*). This is done with 33 mil, 2" wide metal straps placed close to the end of the wall. Lateral or diagonal braces can be screwed and/or welded to the studs on either side (*Figure 32*). This is a type of tension system, meaning it can resist stud buckling through pure tension. This bridging can be helpful for structures that need a lot of mechanical or electrical utilities to run within the wall because it leaves the punchouts open for other uses.

In this system, blocks (cut pieces of studs) should be placed between studs at required intervals to resist rotation. They are typically placed between the first and last stud spaces and at 8'–12' OC. They should be attached to the straps on both sides of the wall.

Continuous blocking can also be installed between each stud as a compression-only system, offering higher brace stiffness (*Figure 33*). They are attached with screws on the outside flange of the studs.

The AISI provides design requirements for steel bracing. Installers should always check code and building requirements.

Nonbearing Walls

Nonbearing walls are walls in the interior of a building that do not need to support structural weight or loads. They are typically intended to divide rooms.

These walls can be braced with cladding on one or both sides using the sheathing braced design. Types of cladding material that can be used include plywood, steel, cementitious or gypsum board, waferboard, and stucco on lath. Gypsum board is the most commonly used form of cladding for bracing (*Figure 34*). One disadvantage to using gypsum board is that if it gets wet, its performance as a brace is significantly reduced. Claddings should be supplemented with steel bridging to align members and ensure structural integrity.

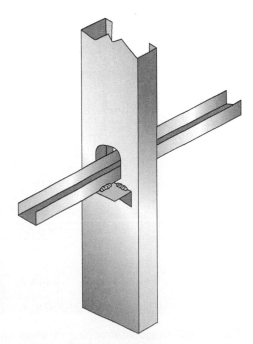

Figure 30 Lateral bridging.

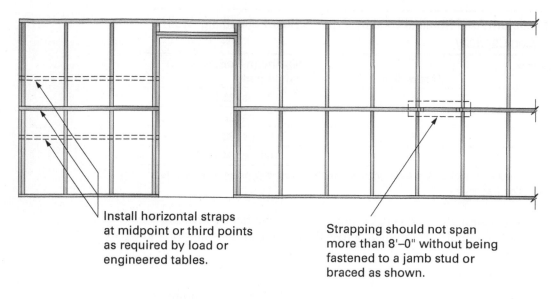

Install horizontal straps
at midpoint or third points
as required by load or
engineered tables.

Strapping should not span
more than 8'–0" without being
fastened to a jamb stud or
braced as shown.

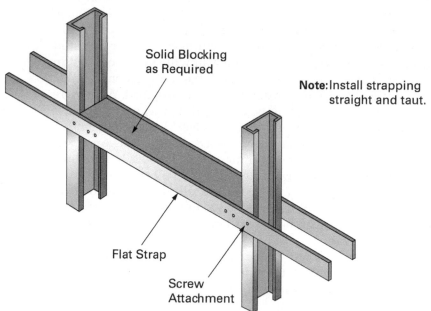

Solid Blocking
as Required

Note: Install strapping
straight and taut.

Flat Strap

Screw
Attachment

Figure 31 Lateral strapping for stud walls.

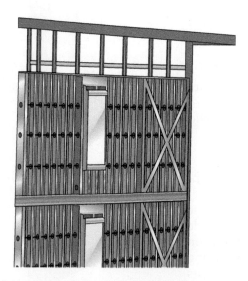

Figure 32 Diagonal strapping for shear walls.

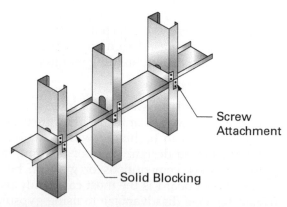

Screw
Attachment

Solid Blocking

Figure 33 Continuous blocking compression-
only system.

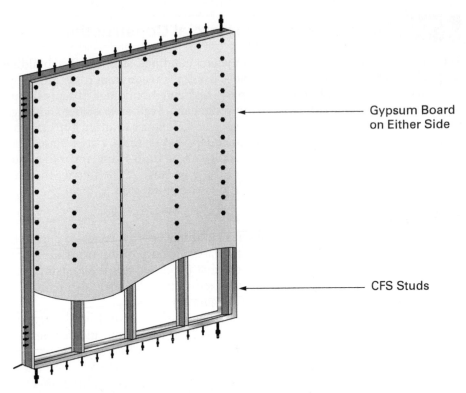

Figure 34 Sheathing braced design with gypsum board.

If there is cladding on one side only, the other side needs to be braced. The minimum interior lateral live load is 5 psf; higher loads may be required by the building specifications.

Steel bracing may also be used instead of cladding. In these instances, a cold-rolled channel can be installed at 4' on center through the punchouts in the studs. Alternatively, other framing products may be used on the face of stud flanges.

Metal Joists and Roof Trusses

In the past, CFS studs were mainly used to frame interior nonbearing walls and partitions in commercial work. Now, CFS can also be used to construct commercial trusses and roofs as well as load-bearing interior and exterior walls. In residential work, an entire house can be framed with CFS studs, joists, and roof trusses.

Steel joists are available in the same sizes as wood joists. Joists can rest directly on concrete or masonry, or they can be attached to a wood sill plate or top plate (*Figure 35*).

One method of installing floor joists in a poured concrete foundation wall is to form slots in the wall to accept the joists (*Figure 36*).

Metal roofs are framed with prefabricated trusses in which the framing members are welded together.

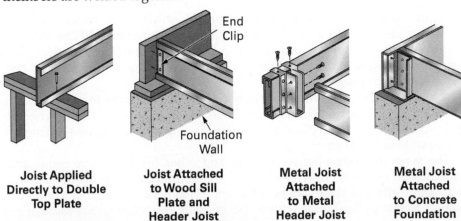

Figure 35 Examples of metal joist installation.

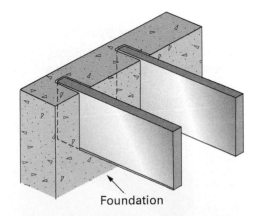

Figure 36 Installing metal floor joists in a slotted foundation wall.

2.3.0   Differences in Commercial Construction

The structural framework of large commercial buildings, such as office buildings, hospitals, apartment buildings, and hotels, is usually made from concrete or structural steel. As larger buildings than many residential buildings, these structures need to meet additional requirements for weight and weather resistance and longevity. Drywall mechanics will need to use specific techniques to attach gypsum board to concrete and masonry. These are discussed in Module 45104, *Drywall Installation*.

Several different types of exterior walls can be built from concrete or steel. Drywall mechanics should be aware of these because of the impact they have on how gypsum board is installed on the inside of the stud.

2.3.1 Concrete and Steel Construction

The exterior finish of large, commercial buildings is often concrete panels that are either prefabricated and raised into place or poured into forms built at the site. Floors are usually made of concrete that is poured at the site using wood, metal, or fiberglass forms. Exterior walls (curtain walls) may also be made of glass in a metal or concrete framework (*Figure 37*).

In some buildings, the framework is made of structural steel. Structural steel is stronger than CFS and is used in construction with greater demands for height, weight, and wind and weather resistance, such as for skyscrapers and parking garages. Because of its strength, structural steel is also used to construct exterior and interior load-bearing walls. Panels fabricated off-site are lifted into place and bolted or welded to the steel (*Figure 38*). Load-bearing walls in residential buildings may also be made of structural steel.

Figure 37 Installing a metal framework.
Source: Lev Kropotov/Shutterstock

Figure 38 Cold formed fabricated exterior wall.
Source: B Brown/Shutterstock

Masonry Veneer System

This masonry wall combines a masonry veneer system with a steel stud backup.

Figure 39 shows the structure of a building in which all the structural framework is made of concrete poured at the site. Each component of the structure requires a different type of form. In this case, the floor and beams were made in a single pour using integrated floor and beam forms, which were removed once the concrete hardened.

In some specialized commercial applications, tilt-up concrete construction is used. In tilt-up construction, the wall panels are usually poured on the concrete floor slab, then tilted into place on the footing using a crane (*Figure 40*). The panels are welded together.

The main difference between tilt-up and other types of large commercial construction is that there is no steel or concrete framework in tilt-up construction. The walls and floor slab bear the entire load. Tilt-up is most common in one- or two-story buildings with a slab at grade (no below-grade foundation). It is popular for warehouses, low-rise offices, churches, and a variety of other commercial and multi-family residential applications. Tilt-up panels of 50' in

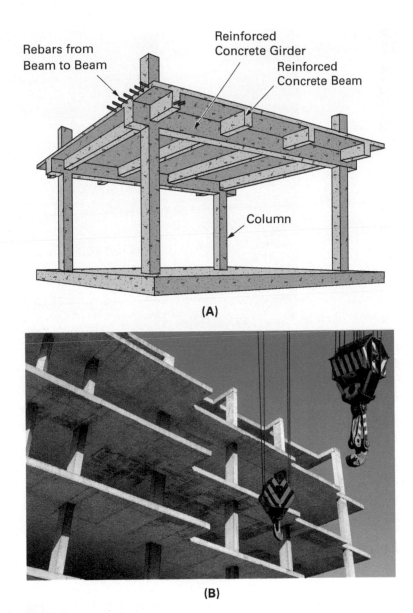

(A)

Rebars from
Beam to Beam

Reinforced
Concrete Girder

Reinforced
Concrete Beam

Column

(B)

Figure 39 Concrete structure.
Source: (B) Eugene Sergeev/Alamy Stock Photo

Figure 40 Tilt-up panel being lifted into place.
Source: IngeBlessas/Shutterstock

Figure 41 ICF wall system.
Source: Alex Hinds/Alamy Stock Photo

height are not uncommon. They typically range from 5" to 8" thick, but thicker walls can be obtained when using lighter-weight concrete.

Insulating Concrete Forms (ICFs) are another type of construction material. In this system, fresh concrete is poured into a cast (cast-in-place), which is set between two layers of insulation, forming a permanent layer of insulation for the building (*Figure 41*). Expanded polystyrene (EPS) or extruded polystyrene (XPS) foam are typically used as the insulation material. ICFs are often used in low-rise residential or commercial buildings. These systems provide additional strength, safety, energy efficiency, and mold and insect resistance.

There are several types of ICF systems. *Flat* systems consist of a solid concrete wall between insulation layers. *Waffle grid* or *grid* systems have a web or waffle pattern of concrete, resulting in areas where the concrete is thicker than others. *Post and beam* systems consist of horizontal and vertical concrete members enclosed in insulation.

2.3.2 CFS Curtain Walls

Curtain walls are walls on the exterior of a building. In general, curtain walls are designed only to accommodate wind loads and transfer those loads back to the structure. They are not designed to carry the gravity loads imparted by the structure itself. There are three types of CFS curtain walls: infill, bypass, and spandrel.

Infill Walls

Infill walls are made of panels that are constructed between the floors of a building's structural frame to support the cladding system. In this system, the steel tracks are built along the lower floor and underside of the upper floor. While they are nonbearing, they can resist wind loads and support their own weight and the weight of cladding. In the past, masonry and timber were used for infill walls, but now CFS with C sections is mainly used for infill walling in steel and concrete framed buildings. When prefabricated as large panels, infill walls may come with the cladding already attached.

CFS infill walls may use a variety of materials for cladding, including the following:

- *Brick* — For brickwork that is supported by the ground or at each floor, the infill wall also provides lateral support with wall ties that are attached with screws through the insulation (*Figure 42*). Taller buildings (over three or four stories) need additional steel angle supports.
- *Aluminum composite panels* — A composite panel of varying thickness, high mechanical strength, and extreme surface flatness.
- *Rain-screen cladding* — A type of cladding that may come in the form of boards, metallic sheets, or tiles attached to horizontal rails.
- *Metallic systems* — These systems, which include composite or sandwich panels, can support terracotta tiles with horizontal rails.

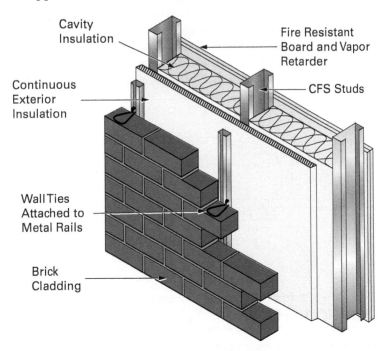

Figure 42 Infill wall with brick cladding.

Wind will cause structures to move, so it is important that the structure is built to allow for movement in a way that maintains structural integrity. Both the exterior cladding and interior gypsum panel must be installed in a way that allows for movement.

Bypass Walls

Bypass walls are sometimes termed *continuous walls* because they extend from the ground up to the roof and can even extend beyond the roof to form a parapet. The framing is outside of the primary structure and a series of clip angles are used to tie the framing back to the primary structure.

For high-rise, commercial constructions, bypass walls may be opaque or transparent glass. When they stretch over the roof, they make the roofline appear to disappear, creating the aesthetically pleasing "glass box" effect.

Spandrel Panels

In the spandrel panel system, the framing is not continuous up through the structure, but only at the floor lines. The panels extend above and below the floor line as needed to meet code or design requirements. The panels also use clip angles to transfer the load to the primary structure.

2.3.3 Concrete Wall Openings

When walls are formed of concrete, openings for doors and windows are made by inserting wooden or metal bucks in the form. Openings for services such as piping and cabling are accommodated with fiber, plastic, or metal tubes inserted into the form.

Openings for doors and windows in CFS walls are built into the design. Pre-punched holes may also be added to the studs when they are being engineered to allow for piping and cabling.

2.3.4 Interior Walls and Partitions for Commercial Buildings

The construction of walls and partitions in commercial applications is driven by the fire and sound-proofing requirements specified in local building codes. In some cases, a frame wall with $\frac{1}{2}$" gypsum board on either side is satisfactory. In other cases, such as the separation between offices and manufacturing space in a factory, it may be necessary to have a concrete block wall combined with fire-resistant gypsum board, along with rigid and/or fiberglass insulation (*Figure 43*). This is especially true if there is any explosion or fire hazard.

While they are not used as often as wood studs in residential construction, steel studs are the standard for framing walls and partitions for commercial construction. Once the studs are installed, one or more layers of gypsum board and insulation are applied. The type and thickness of the board and insulation depend on the fire rating and soundproofing requirements. You will learn more about these in section four.

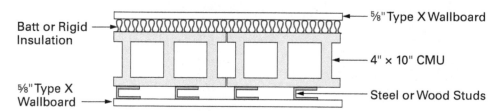

Figure 43 High fire/noise resistance partition.

2.0.0 Section Review

1. Which of the following statements is true?
   a. Wood wall framing is generally done with 2 × 4 studs spaced 10" OC.
   b. Wood wall framing often uses 16" spacing on interior walls.
   c. Some codes permit 36" spacing on exterior walls for one-story buildings.
   d. If 24" spacing is used in a two-story building, the lower floor must be framed with 2 × 4 lumber.

2. Standard metal studs come in widths of $1\frac{5}{8}$" to _____.
   a. 3"
   b. 5"
   c. 6"
   d. 11"

3. The structural framework of large commercial buildings is usually made from which materials?
   a. Wood or CFS
   b. Concrete or structural steel
   c. Masonry or sheet steel
   d. OSB and brick

3.0.0 Ceilings

Objective

Explain ceiling construction processes.

a. Describe residential ceiling construction.
b. Describe commercial ceiling construction.

Performance Tasks

This is a knowledge-based module. There are no Performance Tasks.

Drywall mechanics will primarily be installing materials in the walls and ceilings of a structure and should have general knowledge of the construction of both walls and ceilings.

3.1.0 Residential Ceilings

There are many different options for ceilings in residential construction. Drywall mechanics need to be familiar with the different ceiling options and how they impact drywall installation.

3.1.1 Framing Residential Ceilings

Wood-Framed Ceilings

In wood-framed ceilings, ceiling joists are usually laid across the width of a building at the same positions as the wall studs.

The length of a joist is the distance from the outside edges of the double top plates. The ends of the joists are cut to match the rafter pitch so that the roof sheathing will lay flush on the framing (*Figure 44*). If the joist exceeds the allowable span, two pieces of joist material must be spliced over a bearing wall or partition. *Figure 45* shows two splicing methods. Another method of splicing is to place the two joists on either side of the rafter with a piece of blocking between the joists at the splice.

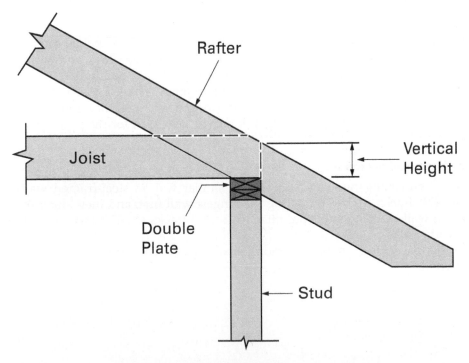

Height of Back of Rafter Above Plate

Figure 44 Cutting joist ends to match the roof pitch.

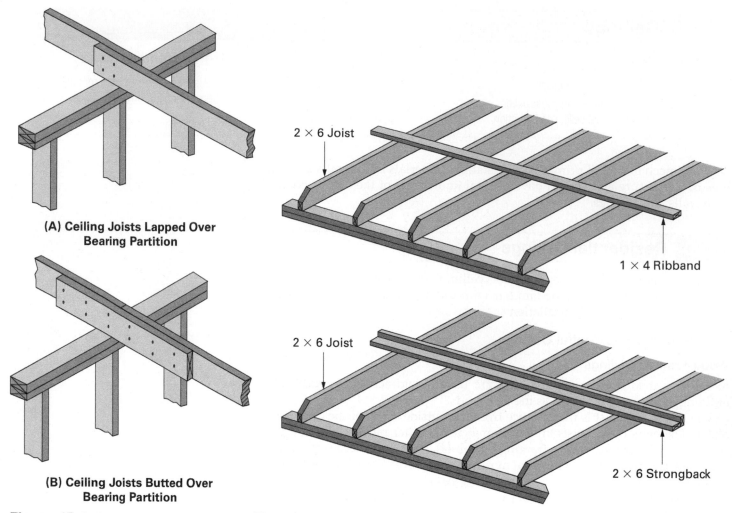

(A) Ceiling Joists Lapped Over Bearing Partition

(B) Ceiling Joists Butted Over Bearing Partition

2 × 6 Joist

1 × 4 Ribband

2 × 6 Joist

2 × 6 Strongback

Figure 45 Spliced ceiling joists.

Figure 46 Reinforcing ceiling joists.

Ribband: A 1 × 4 nailed to ceiling joists to prevent twisting and bowing of the joists.

Strongback: An L-shaped arrangement of lumber used to support ceiling joists and keep them in alignment. In concrete work, it represents the upright support for a form.

If the spacing is the same as that of the wall studs, the joists are nailed directly above the studs. This makes it easier to run ductwork, piping, flues, and wiring above the ceiling. Metal joist hangers can be used in place of nailing.

After the joists are installed, a **ribband** or **strongback** is nailed across them to prevent twisting or bowing (*Figure 46*). The strongback is used for larger spans. In addition to holding the joists in line, it provides support for the joists at the center of the span.

Steel-Framed Ceilings

Steel-framed ceilings are constructed in a similar way to steel-framed walls (*Figure 47*). Tracks are laid out along the longest wall first, and then along the opposite walls. Ceiling studs are attached to the tracks at 16" intervals using pan head screws.

Residential ceilings are not commonly made of CFS. They are typically made with wood.

Figure 47 Steel-framed ceiling.
Source: Dennis Axer/Alamy Stock Photo

3.1.2 Residential Ceiling Design

Of the many types of residential ceiling designs, the most conventional one is the gypsum board design. Applying gypsum boards to the surface of the studs or insulation material and then painting them creates the smooth, seamless look to walls and ceilings that is so signature to residential homes (*Figure 48*). Take a look at the rooms in your own place of residence to see if they were finished with gypsum board.

Other types of residential ceiling designs include the following:

- *Acoustic ceiling tiles* — While acoustic ceiling tiles are often seen in office buildings, they can be found in homes as well, particularly in basements (*Figure 49*). They are created using a suspended ceiling (discussed more in the next section) to hide ductwork and sprinkler systems. They can also help buffer noise from the rooms above or pipes.

Figure 48 Flat gypsum board ceiling.
Source: Paul Maguire/Alamy Stock Photo

Figure 49 Acoustic ceiling tiles.
Source: Star Supawan/Shutterstock

- *Sloped ceiling* — If a building has a triangular roof, the room below likely has a sloped ceiling. This type of ceiling can be finished with gypsum board, but may present a more challenging installation process for drywall mechanics than a simple, flat ceiling.

- *Vaulted ceiling* — This type of ceiling is similar to a sloped ceiling, but it does not need to follow the shape of the roof. These raised ceilings may go even higher than a sloped ceiling to create a more spacious effect in the building's interior. Types of vaulted ceilings include cathedral, domed, barrel, cloister, and shed ceilings (*Figure 50*).

- *Cove ceiling* — These ceilings are shaped like the rounded top of a cave or "cove." They may simply have rounded edges to soften the room, or they may be more completely rounded all the way across. These can be rounded like dome ceilings, giving the room the feeling of more light and space.

- *Coffered ceiling* — This type of ceiling can be found in both residential and commercial buildings (*Figure 51*). They feature square or polygon sunken panels bordered by wooden beams. These can be ornate systems built for aesthetic appeal. Gypsum board may need to be cut in specific dimensions to fit into these types of ceilings.

- *Tray ceiling* — This type of ceiling looks like an upside-down tray. The inside (tray area) of the ceiling extends higher, giving the ceiling more depth, while the border (rim of the tray) is lower and frames the edge of the ceiling.

- *Exposed ceiling* — This type of ceiling features the elements that already exist in the structure, exposing things like beams, joists, and pipes. It is also known as a *raw ceiling*. The beamed ceiling is a similar type of ceiling that mainly involves exposing beams or joists in the design (*Figure 52*). These tend to give the building a more rustic look. Drywall mechanics may need to hang gypsum board around the beams for these ceilings, depending on the design instructions.

Figure 50 Vaulted cathedral ceiling.
Source: Horizon International Images/Alamy Stock Photo

Figure 51 Coffered ceiling.
Source: jimkruger/Getty Images

Figure 52 Beamed ceiling.
Source: Digital Storm/Shutterstock

3.2.0   Commercial Ceilings

Many types of commercial ceilings are in use today as well. Some are similar to or the same as residential ceiling styles.

3.2.1 Suspended Ceilings

Although suspended ceilings are sometimes found in residential applications, they are most commonly used in commercial construction. Suspended ceilings have the following advantages in commercial work:

- They provide excellent noise suppression.
- They provide an area in which horizontal runs of cabling, piping, heating and cooling ducts, and other services can be readily accessed.
- In many commercial buildings, the area between the suspended ceiling and the floor above acts as the return air **plenum** for air conditioning and heating, eliminating the need for some of the sheet metal ductwork.
- The use of suspended ceilings eliminates the need for ceiling framing, as well as the need to box in horizontal runs of ductwork and piping.

There are a wide variety of suspended acoustical ceiling systems. They use the same basic materials, but their appearances are completely different. The focus in this module is on the following systems:

- Exposed grid systems
- Metal panel systems
- Integrated ceiling systems
- Suspended drywall grid ceiling systems
- Special ceiling systems

3.2.2 Exposed Grid Systems

For an exposed grid suspended ceiling, also called a direct-hung system, a light metal grid is hung by wire from the original ceiling or the deck above. Ceiling panels, which are usually 2' × 2' or 2' × 4', are then placed in the frames of the

Plenum: A sealed chamber for moving air under slight pressure at the inlet or outlet of an air conditioning system. In some commercial buildings, the space above a suspended ceiling often acts as a return air plenum.

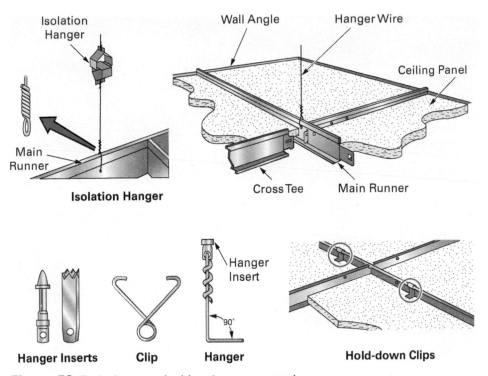

Figure 53 Typical exposed grid system components.

metal grid. Exposed grid systems are constructed using the components and materials shown in *Figure 53* and described as follows:

- *Main runners* — These are the primary support members of the grid system for all types of suspended ceiling systems. They are 12' in length and are usually made in the form of an inverted T. When it is necessary to lengthen the main runners, they can be snapped together with additional runners at the ends.

- *Cross runners (cross ties or cross tees)* — These supports are inserted into the main runners at right angles and spaced an equal distance from each other, forming a complete grid system. They are held in place by either clips or automatic locking devices. Typically, they are either 4' or 2' in length and are usually constructed in the form of an inverted T. Note that 2' cross runners are only required for use with 2' × 2' ceiling panels.

- *Wall angle* — These supports are installed on the walls to support the exposed grid system at the outer edges.

- *Ceiling panels* — These panels are laid in place between the main runners and cross ties to provide an acoustical treatment. Acoustical panels used in suspended ceilings stop sound reflection and reverberation by absorbing sound waves (*Figure 54*). These panels are typically designed with numerous tiny sound traps consisting of drilled or punched holes or fissures, or a combination of both. A wide variety of ceiling panel designs, patterns, colors, facings, and sizes are available, allowing most environmental and appearance demands to be met. Panels are typically made of fiberglass or mineral fiber. Generally, fiberglass panels have a higher sound absorbency than mineral fiber panels. Panel facings are typically made of embossed vinyl and are available in a variety of patterns, such as fissured and pebbled. The specific ceiling panels used must be compatible with the ceiling suspension system, however, because there are variations among manufacturers' standards, and not all panels fit all systems.

- *Hanger inserts and clips* — Many types of fastening devices are used to attach the grid system hangers or wires to the building's horizontal structure above the suspended ceiling. Hanger clips are commonly used to fasten into reinforced concrete with a powder-actuated fastening tool. Clips are used where

Figure 54 Acoustical ceiling panel system.
Source: James Newton-VIEW/Alamy Stock Photo

beams are available and are typically installed over the beam flanges. Then the hanger wires are inserted through the loops in the clips and secured. These devices must be adequate to handle the load.

- *Hangers* — These are attached to the hanger inserts, pins, and clips to support the suspended ceiling's main runners. The hangers can be made of No. 12 wire or heavier rod stock. Ceiling isolation hangers are also available that isolate ceilings from noise traveling through the building structure.

- *Hold-down clips* — These clips are used in some systems to hold the ceiling panels in place. They are used where the ceiling may experience wind loads.

- *Nails, screws* — These fasteners are used to secure the wall angle to the wall. The specific item used depends on the wall construction and material.

Exposed grid ceilings can come in different shapes. A popular shape to use for the grid and panels is a triangle, which can be combined with rectangles in an interesting design (*Figure 55*).

The following are different varieties or variations of exposed grid ceilings.

- *Acoustic ceiling tiles* — This type (as mentioned in the previous section) is also used in residential applications (*Figure 56*).

Figure 55 Triangle exposed grid ceiling.
Source: Sheri Blaney/Stockimo/Alamy Stock Photo

Figure 56 Acoustic ceiling tiles.
Source: Andrew Angelov/Alamy Stock Photo

- *Plaster cast ceilings* — Plaster casts can be made by a manufacturer to add a higher level of design to exposed grid ceilings. The ceiling shown in *Figure 57* features a botanical plaster cast design in an exposed grid system. Plaster casts are also used to create decorative walls.

- *Suspension system* — Along with several other types, this system offers a unique and visually appealing way to use the exposed grid system. It features many exposed grid panels and different shapes extended and layered from the ceiling (*Figure 58*).

Figure 57 Plaster cast exposed grid ceiling.
Source: Hugh Threlfall/Alamy Stock Photo

Figure 58 Suspension system.
Source: Peter Noyce PLBS/Alamy Stock Photo

3.2.3 Metal Panel Systems

The metal panel system is similar to the conventional suspended acoustical ceiling system, but metal panels are used in place of the conventional acoustical panel.

The panels are made of steel (interior or exterior) or aluminum (interior only) and can be painted white, gray, and custom colors. Panels are also available in a variety of surface patterns, such as clusters of circular holes. Tests have indicated that metal panel ceiling systems are effective for sound absorption. They are durable and easily cleaned and disinfected. In addition, the finished ceiling has little or no tendency to have sagging joint lines or drooping corners. The metal panels have a specialized grid system to provide a flush ceiling.

Take care in handling the panels if you must remove them. Use white gloves or rub your hands with cornstarch to keep any perspiration marks from the surface of the panels. If care is not taken, fingerprints will be plainly visible when the units are reinstalled.

When installing or removing panels, follow the manufacturer's instructions carefully. The steps and pieces will vary based on the type of metal panel system (*Figure 59*).

Figure 59 Metal ceiling panel.
Source: David R./Alamy Stock Photo

3.2.4 Integrated Ceiling Systems

As indicated by its name, the integrated ceiling system incorporates the lighting and/or air supply diffusers as part of the overall ceiling system (*Figure 60*).

This system is available in units called *modules*. The common sizes are 30" × 60" and 60" × 60". The dimensions refer to the spacing of the main runners and cross tees.

3.2.5 Suspended Drywall Grid Ceiling Systems

The suspended drywall grid system is used when it is desirable or specified to use a gypsum board finish or a gypsum board backing for an acoustical tile ceiling. This system combines drywall ceiling installation with the exposed tee grid system. Gypsum board is hung on a metal grid system suspended by wires. These can be flat or curved systems that—once installed—have a smooth and seamless surface (*Figure 61*).

Most suspended ceiling system manufacturers have their own proprietary direct-hung drywall suspension system. In the system pictured, various clips are used to attach the gypsum board to the main beams and cross tees.

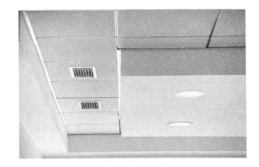

Figure 60 Typical integrated ceiling system.
Source: Bespaliy/Shutterstock

Figure 61 Curved drywall grid system.
Source: hiv360/Shutterstock

Plenum Ceilings

The systems that provide heating and cooling for most commercial buildings are forced-air systems. Blower fans are used to circulate the air. The blower draws air from the space to be conditioned and then forces the air over a heat exchanger, which cools or heats the air. In a cooling system, for example, the air is forced over an evaporator coil that has very cold refrigerant flowing through it. The heat in the air is transferred to the refrigerant, so the air that comes out the other side of the evaporator coil is cold. In homes, the air is delivered to the conditioned space and returned to the air conditioning/heating system through ductwork that is usually made of sheet metal. In commercial buildings with suspended ceilings, the space between the ceiling and the overhead decking is often used as the return air plenum. (A plenum is a sealed chamber at the inlet or outlet of an air handler.) This approach saves money by eliminating about half the cost of materials and labor associated with ductwork.

Anything in the plenum space (electrical or telecommunications cable, for example) must be specifically rated for plenum use in order to meet fire ratings. Plastic sheathing used on standard cables gives off toxic fumes when burned. Plenum-rated cable uses nontoxic sheathing.

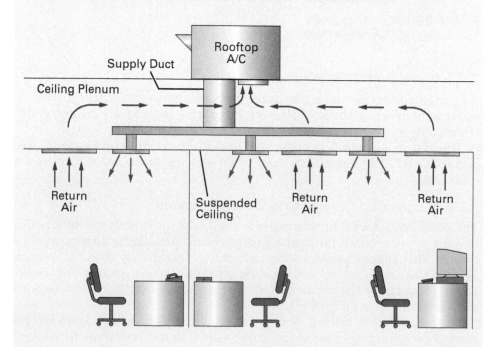

3.2.6 Special Ceiling Systems

Numerous special ceiling systems differ from those covered in this module. A few examples follow:

- *Translucent ceilings* — These ceilings are made up of translucent polycarbonate panels that can generally be attached to a suspension system (*Figure 62*). They brighten up the room with a constant source of muted light.

- *Wood panels or grilles* — Wood panels (or grilles) create beautiful, modern ceilings with the natural feeling of wood (*Figure 63*).

- *Direct surface or direct apply* — Direct surface (also known as *direct apply*) are panels that can be adhered directly to the ceiling surface. These typically attach to exposed ceilings or gypsum ceilings. They can be used to enhance the design or acoustics of the room without having to redo the entire ceiling.

Figure 62 Translucent ceiling.
Source: Vladitto/Shutterstock

Figure 63 Wood panel ceiling.
Source: onurdongel/Getty Images

- *Wood fiber* — These ceilings are made of wood fiber panels in different shapes and designs. They may be direct apply or exposed grid systems.
- *Metal blade* — A metal blade ceiling consists of an exposed grid above a series of metal blades arranged in horizontal rows or a unique pattern (*Figure 64*). These create sleek office ceilings and can be easily attached to main beams or cross tees with attachment clips.

Figure 64 Metal blade ceiling.
Source: Josh Hawley/Getty Images

3.2.7 Seismic Designs

Buildings are prone to extreme damage in areas near fault lines (*Figure 65*). The *IBC*® and other codes have requirements for buildings in these seismic areas to prevent or minimize damage during earthquakes. Builders must be aware of these requirements to ensure they are complying with local and national codes and keeping the building occupants safe.

Figure 65 Building damaged in an earthquake.
Source: fotostory/Shutterstock

Seismic ceilings are designed for areas that are expected to have seismic activity, such as ground shaking from earthquakes. There are six seismic design categories, ranging from Category A—for buildings with minor ground shaking and good soil—to Categories E and F, for buildings near major active faults.

Builders must meet code requirements when constructing ceilings for areas with seismic performance. There are many requirements, which vary by category. For instance, one *IBC*® requirement for buildings in categories D–F is to have 2" wall molding on each side.

Ceiling manufacturers provide different types of ceilings to comply with seismic requirements. For instance, Armstrong's Seismic Rx® Suspension System complies with code requirements for categories C–F. Some of its seismic features include tighter installation, $\frac{7}{8}$" wall molding, and $\frac{3}{8}$" clearance on two unattached walls. Installation and access for seismic ceilings varies from system to system. Installers must follow manufacturer guidelines.

3.2.8 General Guidelines for Accessing Suspended Ceilings

The following are some general guidelines for working with suspended ceilings:

- Contact building maintenance personnel to find out how the ceiling is constructed and how to obtain spare panels in case some of the existing panels get damaged. They should also have the special tools you will need to get access to some types of ceilings. One type of concealed grid system, for example, has a special hook that is used to reach under the panel and release it from the cross member. As previously discussed, pan ceilings require special procedures for removing and installing pans.

- Do not force ceiling panels. Some panels are clipped to the gridwork. If that is the case, you will need to find the panel that is not clipped and start there. A special tool may be needed to release the clips.

- Some ceilings have hinged panels that can be raised or lowered to provide access to the area above.

- Keep your hands clean to avoid staining the ceiling panels. If a panel gets dirty, try cleaning it with a damp sponge or an art gum eraser. Vinyl-faced fiberglass and mylar-faced ceilings can be cleaned with mild detergents or germicidal cleaners.

- Pan ceiling panels require special handling. Wear gloves or rub cornstarch on your hands to prevent the transfer of fingerprints to the panels.

3.0.0 Section Review

1. Which conventional type of residential ceiling design has a smooth, seamless appearance?
   a. Vinyl tile
   b. Gypsum board
   c. Coffered
   d. Exposed

2. When the space above a suspended ceiling is used as the air return for the heating and cooling system, the space is referred to as a _____.
   a. ribband
   b. ceiling duct
   c. plenum
   d. vaulted ceiling

4.0.0 Fire-Rated and Sound-Rated Construction

Performance Tasks	**Objective**
This is a knowledge-based module. There are no Performance Tasks.	Describe fire-rated and sound-rated construction and explain why they are important.

Describe fire-rated and sound-rated construction and explain why they are important.

a. Identify fire and sound rating requirements for walls.

b. Define firestopping and explain how it is accomplished.

c. Describe sound isolation construction and requirements.

Fire resistance: The ability of materials to prevent or retard the passage of excessive heat, hot gases, or flames.

Fire-resistance rating: The time it takes for a building material or assembly to confine a fire and continue to perform its structural function.

Certain walls, floors, and ceilings in a building are rated for their **fire resistance**, as established by building codes. The **fire-resistance rating** is stated in terms of hours, such as one-hour wall or two-hour wall. The rating denotes the length of time an assembly can withstand fire and provide safe evacuation of occupants, as determined under laboratory conditions. The greater the fire-resistance rating, the thicker the wall is likely to be.

The Underwriters Laboratories (UL) tests products and technology for safety. The history of the company can be traced all the way back to the Chicago World's Fair in 1890 when founder William Henry Merrill, Jr., was assessing new fairground construction for fire risks for the Boston Board of Fire Underwriters. Now, the company tests building materials for fire resistance. Builders and designers need to look at the UL ratings on building materials to find the right materials needed to construct code-compliant buildings.

Figure 66 and *Figure 67* show examples of the fire ratings of gypsum board partitions for wood and steel framing as provided by the National Gypsum Company. *ASTM E119* is the testing standard used for determining the fire-resistance rating of partitions, beams, columns, floor-ceiling assemblies, and roof-ceiling assemblies. These are just a selection of examples; different specifications are provided by different organizations. Notice that these also indicate the sound rating in the STC column.

Buildings are also rated for sound. Builders need to meet certain requirements when choosing and installing building elements, including wall, ceiling, and floor materials, doors, windows, insulation, etc. This requires that workers pay close attention to sealing openings to ensure an adequate level of soundproofing.

4.1.0   Fire and Sound Rating Requirements for Walls

Drywall mechanics need to pay close attention to the fire and sound rating requirements for walls. These will affect everything from the type of material to the way it is installed.

4.1.1 Walls Separating Occupancies

In multi-family residential construction, such as apartments and townhouses, the walls and ceilings dividing the occupancies must meet special fire and soundproofing requirements. The code requirements will vary from one location to another and may even vary within areas of a jurisdiction. For example, dwellings in high-risk areas may have stricter standards than those in other areas of the same city or county.

In some cases, codes may require a masonry wall between occupancies. This masonry wall may even be required to penetrate the roof of the building so that if a fire occurs, it is contained within the unit in which it started because it is unable to travel through the walls or across the attic space.

GYPSUM BOARD PARTITIONS – WOOD FRAMING (LOAD-BEARING) – CONTINUED

Item No.		Fire Rating	UL/GA Design	Description	STC	Test No.
10	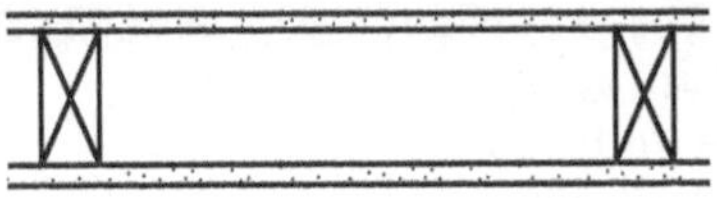	45 Min.	U317	1/2" (12.7 mm) Fire-Shield C Gypsum Board applied vertically to each side of 2 × 4 (38.1 mm × 88.9 mm) wood studs 16" (406 mm) o.c. with 1⅝" (41.3 mm) long. 5d coated nails 7" (178 mm) o.c. Joints staggered on opposite sides.	34	NGC 2161
				Sound rating with 3½" (88.9 mm) glass fiber insulation in stud cavity.	36	NGC 2012051
11	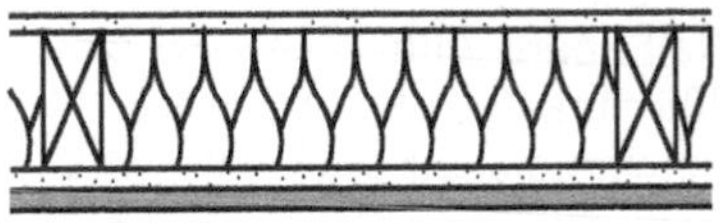	45 Min.	U317	1/2" (12.7 mm) Fire-Shield C Gypsum Board applied vertically to each side of 2 × 4 (38.1 mm × 88.9 mm) wood studs 16" (406 mm) o.c. with 1⅝" (41.3 mm) long. 5d coated nails 7" (178 mm) o.c. ½" (12.7 mm) SoundBreak XP Wall Board applied vertically to one side with 1¼" (31.8 mm) Type S screws 16" (406 mm) o.c. Joints staggered on opposite sides. 3½" (88.9 mm) glass fiber insulation in stud cavity.	43	NGC 2009040
12	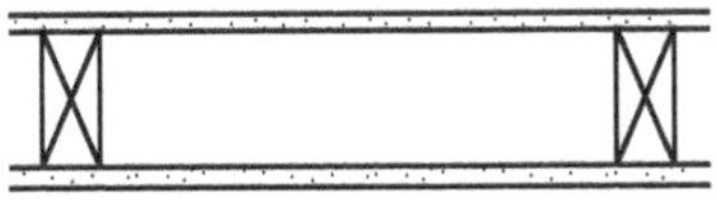	1 Hr.	U305 WP 3605	⅝" (15.9 mm) Fire-Shield Gypsum Board applied vertically or horizontally to each side of 2 × 4 (38.1 mm × 88.9 mm) wood studs 16" (406 mm) o.c. with 1⅞" (47.6 mm) long. 6d coated nails 7" (178 mm) o.c. Joints staggered on opposite sides.	35	NGC 2403
					36	NGC 2008029
				Sound rating with 3½" (88.9 mm) glass fiber insulation in stud cavity.		
13	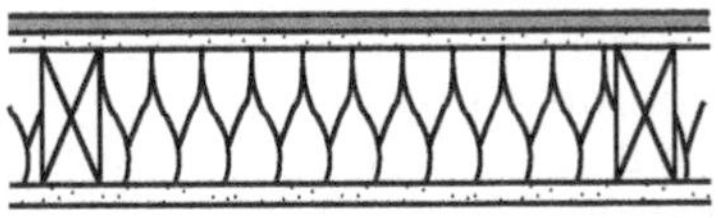	1 Hr.	U305 WP 3605	⅝" (15.9 mm) SoundBreak XP Wall Board applied vertically to one side of 2 × 4 (38.1 mm × 88.9 mm) wood studs 16" (406 mm) o.c. with 1⅞" (47.6 mm) long. 6d coated nails 7" (178 mm) o.c. ⅝" (15.9 mm) Fire-Shield Gypsum Board applied vertically to opposite side of studs with 1⅞" (47.6 mm) long, 6d coated nails 7" (178 mm) o.c. Joints staggered on opposite sides. 3½" (88.9 mm) glass fiber insulation in stud cavity.	42	NGC 2009020
14	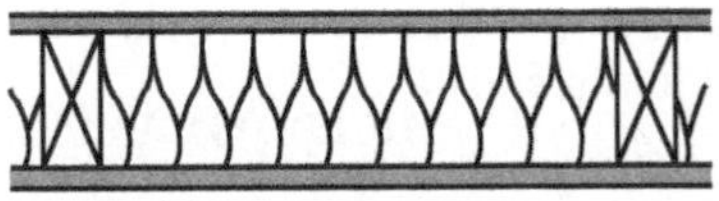	1 Hr.	U305 WP 3605	⅝" (15.9 mm) SoundBreak XP Wall Board applied vertically to each side of 2 × 4 (38.1 mm × 88.9 mm) wood studs 16" (406 mm) o.c. with 1⅞" (47.6 mm) long, 6d coated nails 7" (178 mm) o.c. Joints staggered on opposite sides. 3½" (88.9 mm) glass fiber insulation in stud cavity.	45	NGC 2009019
15	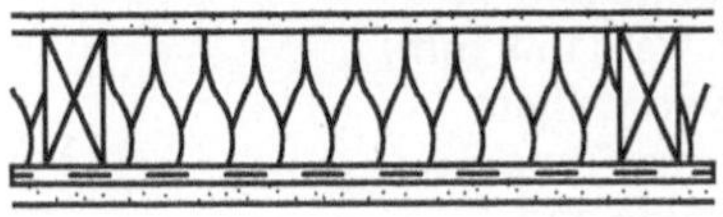	1 Hr.	U305 WP 3249	⅝" (15.9 mm) Fire-Shield Gypsum Board applied vertically to one side of 2 × 4 (38.1 mm × 88.9 mm) wood studs 16" (406 mm) o.c. with 1¼" (31.8 mm) Type W screws 8" (203 mm) o.c. Resilient channels 24" (610 mm) o.c. applied horizontally to opposite side of studs with 1¼" (31.8 mm) Type W screws. ⅝" (15.9 mm) Fire-Shield Gypsum Board applied vertically to channel with 1" (25.4 mm) Type S screws 8" (203 mm) o.c. Joints staggered on opposite sides. 3½" (88.9 mm) glass fiber insulation in stud cavity.	51	NGC 2011071

Figure 66 Specifications for typical fire-rated walls with wood framing.
Source: National Gypsum

GYPSUM BOARD PARTITIONS – STEEL FRAMING (NON-LOAD -BEARING) – CONTINUED

Item No.		Fire Rating	UL/GA Design	Description	STC	Test No.
53	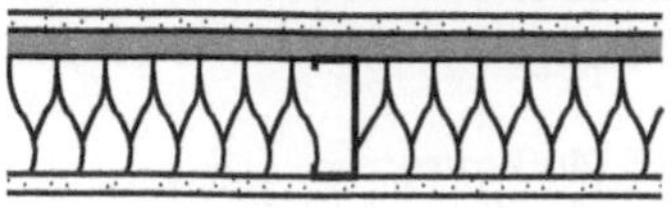	1Hr.	V438 U465	5/8" (15.9 mm) SoundBreak XP Wall Board applied vertically to one side of 35/8" (92.1 mm) steel studs 24" (610 mm) o.c. with 1" (25.4 mm) Type S screws 8" (203 mm) o.c. at vertical joints and 12" (305 mm) o.c. at intermediate studs. 5/8" (15.9 mm) Fire-Shield Gypsum Board applied vertically to opposite side of studs with 1" (25.4 mm) Type S screws 8" (203 mm) o.c. at vertical joints and 12" (305 mm) o.c. at intermediate studs. 5/8" (15.9 mm) Fire-Shield Gypsum Board applied vertically to studs through the SoundBreak XP Wall Board with 15/8" (41.3 mm) Type S screws 12" (305 mm) o.c. 31/2" (88.9 mm) glass fiber insulation in stud cavity.	57	RAL-TL06-334
54	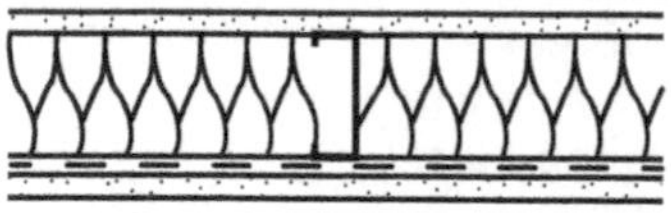	1Hr.	V438 U465	5/8" (15.9 mm) Fire-Shield Gypsum Board applied vertically to one side of 35/8" (92.1 mm) steel studs 24" (610 mm) o.c. with 1" (25.4 mm) Type S screws 8" (203 mm) o.c. at vertical joints and 12" (305 mm) o.c. at intermediate studs. Resilient channels 24" (610 mm) o.c. applied horizontally to opposite side with 1/2" (12.7 mm) Type S-12 screws. 5/8" (15.9 mm) Fire-Shield Gypsum Board applied vertically to channels with 1" (25.4 mm) Type S screws 12" (305 mm) o.c. 31/2" (88.9 mm) glass fiber insulation in stud cavity.	51	NGC 2016017
55	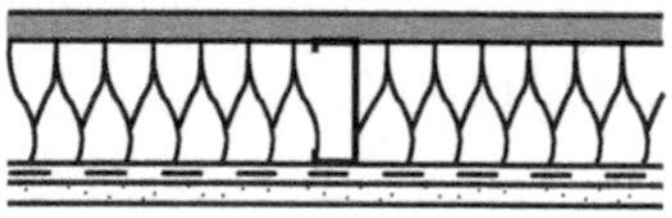	1Hr.	V438 U465	5/8" (15.9 mm) SoundBreak XP Wall Board applied vertically to one side of 35/8" (92.1 mm) steel studs 24" (610 mm) o.c. with 1" (25.4 mm) Type S screws 8" (203 mm) o.c. at vertical joints and 12" (305 mm) o.c. at intermediate studs. Resilient channels 24" (610 mm) o.c. applied horizontally to opposite side with 1/2" (12.7 mm) Type S-12 screws. 5/8" (15.9 mm) Fire-Shield Gypsum Board applied vertically to channels with 1" (25.4 mm) Type S screws 12" (305 mm) o.c. 31/2" (88.9 mm) glass fiber insulation in stud cavity.	56	NGC 2016018
56	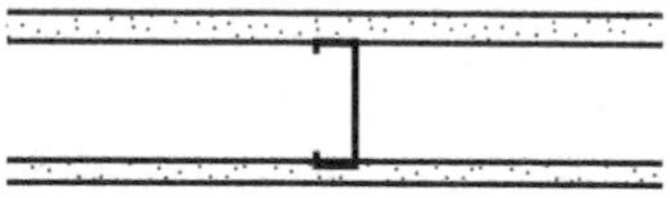	1Hr.	V450 V438 U465	5/8" (15.9 mm) Fire-Shield Gypsum Board applied vertically to both sides of 35/8" (92.1 mm) steel studs 24" (610 mm) o.c. with 11/4" (31.8 mm) Type S screws 8" (203 mm) o.c. at vertical joints and 12" (305 mm) o.c. at intermediate studs.	39	RAL TL05-078
57	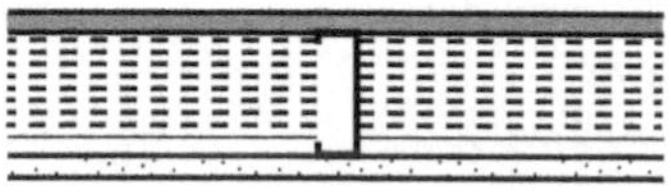	1Hr.	V483	5/8" (15.9 mm) SoundBreak XP Wall Board applied vertically to one side of 35/8" (92.1 mm) steel studs 24" (610 mm) o.c. with 1" (25.4 mm) Type S screws 8" (203 mm) o.c. at vertical joints and 12" (305 mm) o.c. at intermediate studs. 5/8" (15.9 mm) Fire-Shield Gypsum Board applied vertically to opposite side of studs with 1" (25.4 mm) Type S screws 8" (203 mm) o.c. at vertical joints and 12" (305 mm) o.c. at intermediate studs. 3" (76.2 mm) mineral fiber insulation friction fit in stud cavity.	54	RAL TL07-389

Figure 67 Specifications for typical fire-rated walls with steel framing.
Source: National Gypsum

There are many different construction methods for so-called party walls. Each is designed to meet different fire and soundproofing standards. The wall is likely to be more than 3" thick and contain several layers of gypsum board and insulation. A fire-rated wall may abut a non-rated partition or wall. In this case, the rated wall must be carried through to maintain the fire rating. *Figure 68* shows a fire-rated wall with steel studs that abuts a non-rated wall.

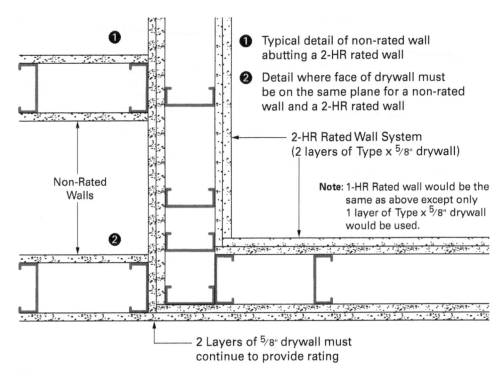

1. Typical detail of non-rated wall abutting a 2-HR rated wall

2. Detail where face of drywall must be on the same plane for a non-rated wall and a 2-HR rated wall

2-HR Rated Wall System (2 layers of Type x ⅝" drywall)

Note: 1-HR Rated wall would be the same as above except only 1 layer of Type x ⅝" drywall would be used.

Non-Rated Walls

2 Layers of ⅝" drywall must continue to provide rating

Figure 68 Example of a fire-rated wall abutting a non-rated wall.

4.1.2 Wall Thickness

The requirements for sound reduction and fire resistance can significantly affect the thickness of a wall. For example, a wall with a high Sound Transmission Class (STC) and fire resistance might have a total thickness of nearly $6\frac{1}{2}$", while a low-rated wall might have a thickness of only $3\frac{1}{2}$" (*Figure 69*).

Party Walls

When gypsum board is used in a party wall, the architect's plans must be followed precisely. If the inspector finds flaws in the construction, a certificate of occupancy will not be issued.

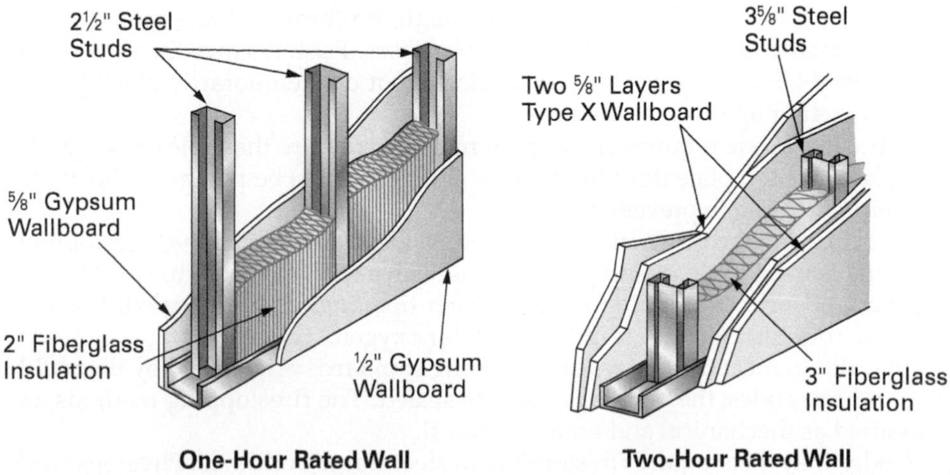

2½" Steel Studs

3⅝" Steel Studs

Two ⅝" Layers Type X Wallboard

⅝" Gypsum Wallboard

2" Fiberglass Insulation

½" Gypsum Wallboard

3" Fiberglass Insulation

One-Hour Rated Wall

Two-Hour Rated Wall

Figure 69 Partition wall examples.

As discussed previously, the fire rating specified by the applicable building code determines the types and amount of material used in a wall or partition. As shown in *Figure 69*, a one-hour rated wall might be made of single sheets of ⅝" gypsum board on wooden or 18-mil CFS studs. A two-hour rated wall requires CFS studs and two layers of fire-resistant gypsum board.

4.2.0 Firestopping

Firestopping: The process of blocking openings in walls, ceilings, and floors with materials or a mechanical device to prevent the passage of fire and smoke.

Firestop: A piece of lumber or fire-resistant material installed in an opening to prevent the passage of fire.

Firestopping means cutting off the air supply so that fire and smoke cannot readily move from one location to another. Firestops are passive systems that are designed to give building occupants time to evacuate the building before the fire spreads. In some areas, local building codes may require firestops. You will hear the term **firestop** used in a few different ways.

Firestopping can be categorized into the following two types:

- *Joint systems* — These systems protect the joints and spaces within fire-rated assemblies. Examples include the top, bottom, and sides of the wall and slab edge.

- *Penetration firestop systems* — These systems ensure the wall rating is maintained when a service, such as an electrical cable or pipe, must pass through a wall. There are two types:
  - *Through-penetration firestop* — These systems protect penetrations passing all the way through a barrier (wall or floor).
  - *Membrane penetration firestop* — These systems protect penetrations passing through only a section of the barrier, such as outlet boxes or sink drains.

Fireblocking is similar to firestopping, but it involves using generic building materials (rather than specially designed materials) to prevent the spread of fire and smoke in building cavities.

4.2.1 Firestopping in Frame Construction

In frame construction, a firestop is a piece of wood or fire-resistant material inserted into an opening such as the space between studs. Wood firestops are short pieces of 2 × 4 blocking (or 2 × 6 pieces if the wall is framed with 2 × 6 lumber) that are nailed between studs (*Figure 70*). This firestop acts as a barrier to block airflow. It does not put out the fire, but it slows the fire's progress.

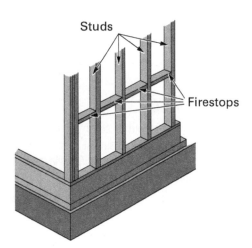

Figure 70 Firestops.

Without firestops, the space between the studs will act like a flue in a chimney, carrying fire rapidly to upper floors. Any holes drilled through the soleplate and top plate create a draft, and air will rush through the space. In a fire, air, smoke, gases, and flames can race through the chimney-like space.

The installation of firestops has two purposes. First, it slows the flow of air, which feeds a fire through the cavity. Second, it can temporarily block flames from traveling up through the cavity.

If the local code requires firestops, it may also require that holes through the soleplate and top plate (for plumbing or electrical runs) be plugged with a firestopping material to prevent airflow.

In commercial construction and some residential applications, firestopping material is used to close wall penetrations such as those created to run conduit, piping, and air conditioning ducts. If such openings are not sealed, fire will travel through the openings in its search for oxygen.

In order to meet the fire-resistance rating standards established by the building and fire codes, the openings must be sealed. The firestopping methods are classified as mechanical and nonmechanical.

Codes may also require firestopping in steel frame buildings. Firestops work a little differently for steel framing than for wood framing. Fire-resistance-rated gypsum boards may be a sufficient firestop in some cases. Manufacturers also provide various firestopping systems that can be attached to steel frames.

4.2.2 Mechanical and Nonmechanical Firestops

Mechanical firestops are fire-resistant devices used to seal the space around the opening where wiring or piping penetrate a wall. There are many different options, such as lighting covers, sleeve kits, cableways, cable transits, pathways, grommets, plugs, and collars.

Nonmechanical firestops are fire-resistant materials, such as caulks and putties, that are used to fill the space around the conduit or piping. You may be required to install various nonmechanical firestopping materials when working with fire-rated walls and floors. Holes or gaps affect the fire rating of a floor or wall. Properly filling these penetrations with firestopping materials maintains the rating.

4.2.3 Fire Safing

Fire safing is a method of firestopping that involves using a "fire safing material" as a firestop. The material is generally mineral wool insulation. For fire safing insulation, it is applied at floor and wall penetrations, construction joints, at the top of wall headers, or at the edge of slabs between floors to create a seal that prevents smoke and fire penetration. It can be used with firestop caulks and sealants.

Fire safing: A type of firestopping that involves using an insulation material (generally mineral wool insulation) as a firestop.

4.3.0 | Sound Isolation Construction

Walls and ceilings must be built for sound containment. To effectively isolate sound, all air leaks and flanking paths, or routes that bypass efforts at sound insulation, must be closed off. Areas where noise can easily travel around—such as around walls, through windows and doors, air ducts, and crawl spaces—must be properly closed off.

Soundproofing needs vary from one use to another and are often based on the amount of privacy required for the intended use. For example, executive and physician offices, high-rise condos, hospitals, and houses for the elderly may require more privacy than general offices.

4.3.1 Sound Ratings

A building may need to meet several sound ratings to pass applicable code requirements.

Buildings must meet an STC rating. The **Sound Transmission Class (STC)** measures how well the interior building assemblies (walls, ceilings, floors, windows, and doors) reduce airborne sound transmission. The higher the STC rating, the better the sound containment. The *IRC*® requires that wall, floor, and ceiling assemblies meet a minimum STC rating of 45 when tested using the ASTM E90 sound transmission loss test. STC ratings can be negatively impacted by hairline cracks or other openings.

Sound Transmission Class (STC): Measures how well the interior building assemblies (walls, ceilings, floors, windows, and doors) reduce airborne sound transmission.

The **Impact Insulation Class (IIC)** measures impact-generated noise created through the floor, ceiling, or roof that passes into the area below. Impacts are caused by footsteps, chair or table scrapes, moving furniture, falling objects, and weather. This is an especially important rating for multi-unit buildings where the impact noise of the unit above can greatly affect the unit below. The IRC® requires that floor/ceiling assemblies between dwelling units or between a dwelling unit and a public or service area within the structure meet a minimum IIC rating of 45 when tested using the ASTM E90 sound transmission loss test.

Impact Insulation Class (IIC): A sound rating that measures impact-generated noise created through the floor, ceiling, or roof that passes into the area below.

The **Outdoor/Indoor Transmission Class (OITC)** measures how well the building assemblies reduce airborne noise traveling from outside to the inside of a building. It is typically used to rate façade elements, such as exterior walls, windows, doors, and roofs. OITC is similar to STC, but was developed to measure noise from outside transportation sources while the STC was developed to focus more on inside noises. It can be especially helpful for buildings that are close to airports, railroads,

Outdoor/Indoor Transmission Class (OITC): A sound rating that measures how well the building assemblies reduce airborne noise traveling from outside to the inside of a building.

highways, and busy streets. The OITC is not currently required by the *IRC*®, but it is still often required by building designers in construction documents.

It is particularly important to know sound requirements when choosing or installing drywall and insulation materials because these will have a big impact on a building's sound rating.

4.3.2 Soundproofing Materials

Soundproofing will be enabled by everything in the wall assembly including the studs, sheathing, insulation, and finishing materials. There are types of materials that are specifically designed to deliver noise control. For instance, **Sound Attenuation Batts (SABs)** are flexible fiberglass insulation batts that are meant to control noise in metal stud wall cavities in interior partitions.

Acoustical caulk (or sound caulk) can be used to seal gaps on the perimeter of sound control walls, floors, and ceilings. It can also be used to seal holes or gaps around penetrations, such as electrical outlets and wires. It is more effective for noise control than regular caulk because it is more durable and does not harden or crack, giving it permanent flexibility.

Sound Attenuation Batts (SABs): Flexible fiberglass insulation batts designed to control noise in metal stud wall cavities in interior partitions.

Acoustical caulk: Caulk used to seal gaps on the perimeter of sound control walls, floors, and ceilings.

4.0.0 Section Review

1. In multi-family residential construction, the walls and ceilings dividing the occupancies must meet special soundproofing and _____ requirements.
   a. fire-resistance rating
   b. support
   c. air-sealing
   d. energy conservation

2. Which of the following statements about firestops is true?
   a. They insulate the walls in a building to prevent heat from traveling out.
   b. They enable the space around wall penetrations to stay open for ventilation.
   c. They stop the flow of air completely, preventing fire from feeding through a cavity.
   d. They can temporarily block flames from traveling up through the cavity.

3. The STC rating indicates _____.
   a. how airtight a structure's duct system is
   b. the thickness of gypsum boards installed
   c. the fire resistance of walls
   d. how well a structure contains sound transmission

Module 45102   Review Questions

1. The popularity of using wood construction globally has gone down over the years because _____.
   a. wood is eco-friendly
   b. wood adds to pollution and climate change
   c. wood can be recycled after use
   d. wood is inexpensive and easy to manufacture

2. Light framing (LF) members used in framework for residential construction are _____.
   a. 1½" thick and 2" wide or wider
   b. 2" to 4" thick and 6" wide or wider
   c. 2" to 4" thick and 2" to 4" wide
   d. 5" and thicker by 8" and wider

3. Wood made from gluing together layers of veneers is known as _____.
   a. plywood
   b. particleboard
   c. oriented strand board
   d. engineered wood

4. The web of a wood I-beam is made of _____ or plywood.
   a. steel
   b. HDO
   c. OSB
   d. LVL

5. Gypsum core board is manufactured in what thickness?
   a. ⅜ inch
   b. ½ inch
   c. ⅝ inch
   d. 1 inch

6. What does Type X gypsum board indicate?
   a. Mold resistance
   b. Foil backing
   c. Improved fire resistance
   d. Water resistance

7. Which type of gypsum board is used for outdoor ceilings such as carports?
   a. Abuse Resistant gypsum board
   b. Gypsum core board
   c. Gypsum shaftliner
   d. Exterior gypsum soffit board

8. Once concrete has hardened and gained its structural strength, it is known as _____.
   a. green concrete
   b. plastic concrete
   c. portland cement
   d. cured concrete

9. What is used to distribute the weight of a window or door opening?
   a. Header
   b. Trimmer stud
   c. Soleplate
   d. Double top plate

10. When CFS studs are used for drywall framing, the channel stock comes in two grades _____.
    a. non-structural and structural
    b. non-structural and track
    c. structural and track
    d. track and steel

11. The edge of the stud that connects to the web on either side and the area where the drywall or other enclosing material is applied to the stud is known as a _____.
    a. slap stud
    b. knockout
    c. flange
    d. track

12. What is the maximum OC spacing for sidewalls with a multi-ply gypsum board base thickness of ½" or ⅝"?
    a. 12"
    b. 16"
    c. 24"
    d. 28"

13. What is the minimum support recommended for load-bearing steel walls?
    a. Plywood cladding
    b. Gypsum board sheathing
    c. Lateral bracing
    d. Diagonal bracing

14. Which type of construction is made up of concrete cast between two layers of insulation?
    a. Brick infill walls
    b. CFS sheets
    c. Tilt-up panels
    d. ICFs

15. Which statement is true about bypass walls?
    a. The framing is outside of the primary structure and a series of clip angles are used to tie the framing back to the primary structure.
    b. The framing is not continuous up through the structure; only at the floor lines.
    c. They are constructed between the floors of a building's structural frame to support the cladding system.
    d. When prefabricated as large panels, they can be attached to brick, insulated render, rain-screen cladding, or metallic systems.

16. In wood-framed ceilings, the length of a joist is _____.
    a. the distance from the inside edges of the double top plates
    b. the distance from the outside edges of the double top plates.
    c. the same distance as the rafters
    d. 16" on center

17. Which of the following statements about wood-framed ceilings is true?
    a. Ceiling joists are usually laid across the width of a building at the same position as the wall studs.
    b. The ends of the joists are cut the opposite of the rafter pitch.
    c. If the joist exceeds the allowable span, three pieces of joist material must be spliced over a bearing wall.
    d. If the spacing is the same as that of the wall studs, the joists are nailed directly below the studs.

18. The _____ ceiling features square or polygon sunken panels bordered by wooden beams.
    a. cove
    b. dome
    c. coffered
    d. vaulted

19. What is the primary support member of grid systems for all types of suspended ceilings?
    a. Cross tees
    b. Main runners
    c. Cross runners
    d. Hanger clips

20. Which type of suspended ceiling system incorporates lighting and/or air supply diffusers into the overall system?
    a. Metal panel system
    b. Integrated ceiling system
    c. Drywall grid ceiling system
    d. Tray ceiling system

21. _____ systems brighten up the room with a constant source of muted light.
    a. Metal blade
    b. Direct surface
    c. Wood fiber
    d. Translucent ceiling

22. Which type of ceiling is specially designed for areas with earthquakes?
    a. Wood grille
    b. Metal blade
    c. Translucent
    d. Seismic

23. Fire-resistance ratings are measured in _____.
    a. hours
    b. inches (of wall thickness)
    c. minutes
    d. millimeters (of wall thickness)

24. Firestops are _____ systems that are designed to give building occupants time to evacuate the building before the fire spreads.
    a. active
    b. passive
    c. impenetrable
    d. cutoff

25. _____ ensure that wall rating is maintained when a service, such as electrical cable or pipe, must pass through a wall.
    a. CFS studs
    b. Penetration firestop systems
    c. Joint systems
    d. Party walls

26. Short pieces of 2 × 4 blocking nailed between studs are required in some areas to _____.
    a. support window sills
    b. slow the spread of fire
    c. reinforce the wall
    d. support electrical cables

27. _____ are fire-resistant materials, such as caulks and putties, that are used to fill the space around conduit or piping where it penetrates a wall.
    a. Joint systems
    b. Mechanical firestops
    c. UL rated steels
    d. Nonmechanical firestops

28. Which method of firestopping generally involves using mineral wool insulation?
    a. Blocking
    b. Fire safing
    c. Mechanical firestops
    d. Nonmechanical firestops

29. The _____ measures impact-generated noise, such as footsteps and moving furniture, created through the floor, ceiling, or roof that passes into the area below.
    a. outdoor/indoor transmission class (OITC)
    b. sound transmission class (STC)
    c. impact insulation class (IIC)
    d. sound containment class (SCC)

30. Which sound rating helps measure outdoor noise for buildings around airports and railroads?
    a. IIC
    b. OITC
    c. STC
    d. ASTM E90

Answers to Odd-Numbered Module Review Questions are found in *Appendix A*.

Answers to Section Review Questions

Answer	Section	Objective
Section One		
1. a	1.0.0	1a
2. c	1.2.1; *Table 2*	1b
3. b	1.3.1	1c
4. d	1.4.1	1d
Section Two		
1. b	2.1.0	2a
2. c	2.2.1	2b
3. b	2.3.0	2c
Section Three		
1. b	3.1.2	3a
2. c	3.2.1	3b
Section Four		
1. a	4.1.1	4a
2. d	4.2.1	4b
3. d	4.3.1	4c

Thermal and Moisture Protection

Source: Pat Canova/Alamy Stock Photo

Objectives

Successful completion of this module prepares you to do the following:

1. Describe insulation and explain why it is used in buildings.
   a. Explain what insulation does.
   b. Describe insulation requirements.
2. Describe insulation materials and types.
   a. Define flexible insulation and explain how it is used.
   b. Define loose-fill insulation and explain how it is used.
   c. Define rigid or semi-rigid insulation and explain how it is used.
   d. Define reflective and spray insulation and lightweight aggregates.
3. Explain how to install different types of insulation.
   a. Describe how to install flexible insulation.
   b. Describe how to install loose-fill insulation.
   c. Describe how to install rigid or semi-rigid insulation.
   d. Describe how to install foam insulation.
4. Explain how moisture control is accomplished in the construction of a building.
   a. Explain how to design a structure to control moisture.
   b. Describe how ventilation and vapor retarders are used for moisture control and explain how they are installed.
5. Explain how water and air infiltration control are accomplished in the construction of a building.
   a. Describe how water-resistive barriers are used to resist liquid water.
   b. Describe how air barriers are used and installed.
   c. Explain how to protect openings, penetrations, and joints.

Performance Tasks

Under supervision, you should be able to do the following:

1. Install blanket insulation in a wall.
2. Install a vapor retarder on a wall.
3. Install water-resistive barriers and materials.

Digital Resources for Drywall

Scan this code using the camera on your phone or mobile device to view the digital resources related to this craft.

Overview

A properly insulated building is comfortable to live or work in and economical to heat and cool. Without insulation, warm air will escape the building in cold weather, causing the heating system to operate constantly. This results in an increased use of energy. In hot weather, the air conditioning system will have to work harder, producing the same result. Proper insulation in walls, floors, and roof decks will minimize this problem. Vapor retarders must also be used to prevent moisture from penetrating the building. Moisture can cause serious problems, including wood decay and mold growth. A skilled craftworker knows how to select and install insulating materials and vapor retarders.

NCCER Industry-Recognized Credentials

If you are training through an NCCER-accredited sponsor, you may be eligible for credentials from NCCER. The ID number for this module is 45103. Note that this module may have been used in other NCCER curricula and may apply to other level completions. Contact NCCER at 1.888.622.3720 or go to **www.nccer.org** for more information.

You can also show off your industry-recognized credentials online with NCCER's digital credentials. Transform your knowledge, skills, and achievements into credentials that you can share across social media platforms, send to your network, and add to your resume. For more information, visit **www.nccer.org**.

1.0.0 Insulation

Performance Tasks

There are no Performance Tasks in this section.

Objective

Describe insulation and explain why it is used in buildings.
  a. Explain what insulation does.
  b. Describe insulation requirements.

Four important considerations for the construction of any building include thermal insulation, moisture control and ventilation, waterproofing, and air infiltration control. This module covers these areas and presents materials and procedures that can be applied to ensure effective installations.

When insulating a structure, you need to pay close attention to these factors. Drywall mechanics need to control the moisture migration as well as the movement of heat through a wall. This is a requirement mandated by code. You will need to ensure that you are complying with all relevant codes, such as the *International Energy Conservation Code*® (*IECC*®), *International Building Code*® (*IBC*®), and *International Residential Code*® (*IRC*®), as well as applicable local codes.

1.1.0 What Is Insulation?

Insulation reduces the transfer of heat through a building. Most materials used in construction have some insulating value. Air is an excellent insulator if it is confined to very small spaces and kept very still. Manufactured insulation material is designed to trap a large amount of air in many tiny spaces to resist the transfer of heat and sound. Double-pane and triple-pane windows also use this method to reduce heat loss.

The amount of insulation in a building directly affects heating and cooling costs. It also affects the value of the building. Some jurisdictions require that a permanent certificate be placed on the building's electrical box stating the insulative properties of material used in the structure. When required, this certificate is completed by the builders or designer.

There are two overall categories of insulation—cavity and continuous. **Cavity insulation** is insulating material that is placed between framing members

Cavity insulation: Insulating materials that are located between framing members.

Figure 1 Cavity insulation.
Source: ScotStock/Alamy Stock Photo

(*Figure 1*). When you think of insulation, the first image that comes to mind is probably fluffy pink fiberglass blankets tucked between wood studs. That is an example of cavity insulation.

Continuous insulation is insulation that runs over a building's structural members continuously (*Figure 2*). It does not have thermal bridges except for fasteners and service openings. It is integral to the opaque surface of a building envelope and can be installed in the interior and exterior.

It is important to know how to use and install both types of insulation to meet code insulation requirements.

Continuous insulation: Insulation that runs over a building's structural members seamlessly without breaks or gaps.

Figure 2 Continuous insulation.
Source: Sergii Petruk/Alamy Stock Photo

Source: stockcreations/Shutterstock

Types of Insulation

There are many different types of insulation. Each type has specific applications for which it is best suited. The following examples are discussed in this module:

- Fiberglass insulation
- Loose-fill insulation
- Rigid or semi-rigid insulation
- Reflective (or foil-faced) insulation
- Sprayed-in-place insulation
- Lightweight aggregates
- Mineral wool insulation
- Kraft-faced insulation

1.1.1 Thermal Resistance

Drywall mechanics must know how thermal resistance works so they can understand the importance of installing insulation properly. They must also be able to follow the requirements of applicable codes, which dictate how much thermal resistance a building should have.

A law of physics states that heat will always flow (or conduct) through any material or gas from a higher temperature area to a lower temperature area. That means that in a building, heat will naturally flow from a warm area into a cool area. Have you ever wondered why the top floor of a house is the warmest on a cold day? That's because the interior heating is moving up to the sky above the ceiling where the air is coldest. Trying to keep a single room warm is a battle between the heat's natural tendency to escape into a cooler area and the protections builders have put in place to prevent its movement. Insulation is essential to keeping a building warm and preventing heat from escaping into the cooler air outside.

Thermal resistance is a measure of how a material will resist the flow of heat energy. Different types of materials have better thermal resistance than others, making them more conducive to insulation.

A material or object's thermal resistance—or ability to resist **heat conduction**—is measured as its **R-value**. As a rule, the higher the R-value, the greater the effectiveness of the insulation. (These requirements are outlined later in the module.) Drywall mechanics must know how to find R-values so they can meet these requirements.

R-value is expressed as:

$$R = 1/k \text{ or } 1/C$$

Where:

k = amount of heat in British thermal units (Btu) transferred in one hour through 1 ft^2 of a material that is 1" thick and has a temperature difference between its surfaces of 1°F; also called the *coefficient of thermal conductivity*

C = conductance of a material, regardless of its thickness; the amount of heat in Btus that will flow through a material in one hour per ft^2 of surface with 1°F of temperature difference

R = thermal resistance; the reciprocal (opposite) of conductivity or conductance

The higher the R-value, the lower the conductive heat transfer. *Table 1* shows the R-values of various common building materials, including some common insulating materials.

Thermal resistance: A measure of how a material will resist the flow of heat energy.

Heat conduction: The process by which heat is transferred through a material, which is caused by a difference in temperature between two areas.

R-value: A measure of an object or material's amount of thermal resistance.

TABLE 1 Approximate R-Values of Common Materials

Material	Thickness	R-Value °F/ft^2/hr in Btus
Air film and spaces		
Air space bound by ordinary materials	½" to 4"	1.00
Masonry units		
Concrete masonry unit (CMU)	8"	1.11
Concrete masonry unit (CMU)	12"	1.28
Common brick	4"	0.80
Sheathing materials		
Fiberglass	1"	4.00
Fiberboard sheathing	½"	1.32
Plywood	½"	0.63
Extruded polystyrene (EPS)	1"	5.00
Foil-faced polyisocyanurate (ISO)	1"	7.20
Siding materials		
Bevel-lapped wood siding	½" × 8"	0.80
Hardboard	½"	0.34
Plywood	¾"	0.94
Brick	4"	0.44
Aluminum, steel, vinyl (hollow backed)		0.61
Stucco	1"	0.20
Common insulating materials		
Fiberglass batts	3½"	11.00
Fiberglass blown (attic)	1"	2.12–4.30
Rigid fiberglass (less than 4lb/ft^3)	1"	4.00
Mineral wool batt	1"	3.14–4.00
Vermiculite	1"	2.13
Polyurethane foam	1"	5.88–6.25
Expanded polystyrene foam	1"	4.00
Interior finish materials		
Gypsum board	½"	0.39

Source: https://www.archtoolbox.com/r-values/

Think About It

Calculating the Effective R-Value

To calculate the effective R-value of an entire assembly, add together the individual R-values of each of the unbroken layers of materials in the assembly.

The total heat transmission through a wall, roof, or floor of a structure in Btu per ft^2 per hour with a 1°F temperature difference is called the total heat transmission or **U-factor**. Whereas the R-value takes into account the individual insulating materials, the U-factor considers the entire assembly. This includes the thermal bridging of framing and studs.

The U-factor is the inverse of the R-value. It is expressed as follows:

$$U = \frac{1}{R_1 + R_2 + \cdots + R_N}$$

U-factor: A measure of the total heat transmission through a wall, roof, or floor of a structure.

Air Versus Inert Gas

The air trapped between the panes of double-pane and triple-pane windows makes a good insulator, but there are better alternatives. Inert gases that block more heat than air create a stronger barrier. That's why some window manufacturers fill the space between the panes with argon—an inert, nontoxic, nonflammable gas.

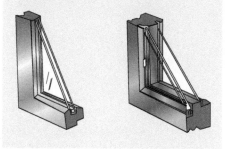

Where:

$R_1 + R_2 + \ldots R_N$ represents the sum of the individual R-values for the materials that make up the thickness of the wall, roof, or floor. As shown in *Table 2*, the whole wall R-value is significantly influenced by the percentage of windows and doors in the wall and their heat transmission.

TABLE 2 Example Wall Systems Based on the U-Factor Approach

2 × 6 Studs Spacing	Cavity Insulation R	% Wall	Continuous Insulation R	% Wall	Windows/ Doors R	% Wall	Whole Wall R-Total
16	21	63	0	0	3	15	8.9
24	21	71	0	0	3	15	9.6
16	21	63	6	85	3	15	11.2
24	21	71	9	85	3	15	12.2
16	21	63	6	85	5	15	14.1
24	21	71	9	85	5	15	18.6

The lower the U-factor, the lower the heat transmission.

While the R-values provide a convenient measure to compare heat loss or gain, the total U-factor for a structure is used in the calculations for sizing the structure's heating and cooling equipment (*Figure 3* and *Table 3*).

TABLE 3 R-Values of Typical Wall Construction

2 × 4 Stud Wall with Rigid Board	
Type	**R-Value**
Air films*	1.00
¾" wood exterior siding	0.93
1" polystyrene rigid board	5.0
3½" batt or blanket insulation	11.0
Vapor retarder	0.0
½" gypsum board	0.39
Total Effective R-Value	18.32
2 × 6 Insulated Stud Wall	
Type	**R-Value**
Air films*	1.00
¾" wood exterior siding	0.93
3" expanded polystyrene foam	12.0
3" rigid fiberglass	12.0
Vapor retarder	0.0
½" gypsum board	0.39
Total Effective R-Value	26.32

*Stagnant air film that forms on any surface

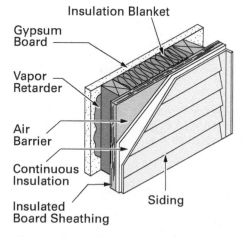

Insulation Blanket
Gypsum Board
Vapor Retarder
Air Barrier
Continuous Insulation
Insulated Board Sheathing
Siding

Figure 3 Typical wall construction.

By doubling the R-value of a wall or roof, the conductive heat loss or gain can theoretically be reduced by half. However, it is important to note that as insulation thicknesses are increased, the heat transmission (U-factor) is decreased, but

Wood Framing Versus Steel Framing

Many modern constructions use steel framing. Steel can be a stronger, more durable material than wood that is more resistant to water, mold, air, and fire. One drawback of steel framing is that it conducts more heat than wood. *ASHRAE®* estimates that the effectiveness of a steel-framed building's insulation can be reduced. This leads to increased energy costs and consumption while detracting from the interior comfort of the building. Exterior continuous insulation is a necessity for steel framed buildings to counteract this effect.

not in a direct relationship. Increases of insulation will continue to decrease heat loss, but at lower and lower percentages. At some point, it becomes economically useless to add more insulation. The same is true for double-, triple-, and quadruple-pane windows. It must also be noted that conductive heat loss or gain does not include heat gains or losses due to air leaks or **radiation** through windows or other openings.

1.2.0 Insulation Requirements

Increasing energy costs and mandated government energy conservation have resulted in much higher R-value requirements for new construction. While building code and design standards for insulation have been traditionally based on average low-temperature zones and charts based on the range of low temperatures expected, the requirements are constantly changing. The *IECC®* recommends insulation values be based on **climate zone**, which is determined by local temperature and humidity levels (*Figure 4* and *Table 4*).

Continuous insulation may be needed to ensure the highest level of insulation and meet code requirements. The code often gives the option of using cavity or continuous insulation (or a combination of both) with different required R-values. The R-values needed for cavity insulation are generally higher to account for the additional protection needed to counter the effects of thermal bridging. In cooler climates, continuous insulation is often required in addition to cavity insulation, as shown in *Table 4*.

Note that *Table 4* shows only a sample of the information available in Table C402.1.3 in Chapter 4 of the *IECC®* Commercial Provisions. The code also includes information on different types of roofs, walls, floors, and opaque doors.

Did You Know?
Over-Insulating

Installing excess insulation wastes money and may cause other problems. If a building is over-insulated and lacks sufficient ventilation and water barrier protection, moisture can collect inside. This promotes the growth of mold and fungus. It is even possible that cancer-causing radon gas could be trapped in the building, accumulating over time to dangerous levels.

Radiation: Energy emitted from a source in electromagnetic waves or subatomic particles.

Climate zone: A geographical region based on climate criteria, as determined by the *IECC®*.

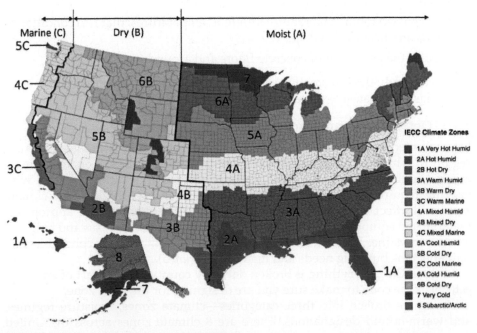

Figure 4 Climate zones in the United States.
Source: Office of Energy Efficiency & Renewable Energy 2021 IECC – International Energy Conservation Code

TABLE 4 Minimum R-Values of Commercial Insulation Requirements

Climate Zone	Floors: Mass		Walls: Wood-Framed		Roofs: Attic and Other	
	All Other	Group R	All Other	Group R	All Other	Group R
1	NR	NR	R-13 + R-3.8ci or R-20	R-13 + R-3.8ci or R-20	R-38	R-38
2	R-6.3ci	R-8.3ci	R-13 + R-3.8ci or R-20	R-13 + R-3.8ci or R-20	R-38	R-38
3	R-10ci	R-10ci	R-13 + R-3.8ci or R-20	R-13 + R-3.8ci or R-20	R-38	R-38
4 except Marine	R-14.6ci	R-16.7ci	R-13 + R-3.8ci or R-20	R-13 + R-3.8ci or R-20	R-49	R-49
5 and Marine 4	R-14.6ci	R-16.7ci	R-13 + R-7.5ci or R-20 + R3.8ci	R-13 + R-7.5ci or R-20 + R-3.8ci	R-49	R-49
6	R-16.7ci	R-16.7ci	R-13 + R-7.5ci or R-20 + R-3.8ci	R-13 + R-7.5ci or R-20 + R-3.8ci	R-49	R-49
7	R-20.9ci	R-20.9ci	R-13 + R-7.5ci or R-20 + R-3.8ci	R-13 + R-7.5ci or R-20 + R-3.8ci	R-60	R-60
8	R-23ci	R-23ci	R-13 + R-18.8ci	R-13 + R-18.8ci	R-60	R-60

ci = Continuous insulation; NR = No Requirement
Group R = Residential – A building or structure intended for sleeping purposes

In general, insulation must be installed where any exterior surface of a structure is exposed to a thermal difference relative to its internal surface. These areas are as follows:

- Roofs
- Above ceilings
- In exterior walls
- Beneath floors over crawl spaces
- Between two rooms that require different temperatures; that is, a conditioned versus an unconditioned space
- Around the perimeter of concrete floors and around foundations

As you study the information in *Table 4*, you will notice that the ceiling insulation has the greatest R-value.

1.2.1 Defining Climate Zones

Drywall mechanics must consider the climate zone. This is the geographical region in which a construction project is taking place, based on the climate criteria outlined in the *IECC*®. Regions are divided into different zones with varying requirements due to climate differences in each region. The climate zone will affect everything from the scope of the work to the appropriate material and equipment to use. Consider building a house in Texas and building a house in Maine. These states have extremely different weather conditions that will affect the building needs and materials (*Figure 5*).

In the *IECC*®, everything is broken down by county. At the start of any project, check the code to make sure you are designing for the right zone.

Zones are broken into three categories—climate zones, moisture regimes, and warm-humid designations. There are 8 climate zones across the United States. For example, as shown in *Figure 4*, the northern part of the state of Illinois is in Zone 5. Cook County is in Zone 5A. The A, B, and C refer to the moisture regimes (A: Moist, B: Dry, and C: Marine). In the *IECC*®, if the county is marked

Figure 5 Different seasons and climates.
Source: Artur Debat/Getty Images

with an asterisk, it is a warm-humid location. Cook county is not marked with an asterisk. Therefore, if you were installing insulation in a house in Chicago (in Cook County), you would need to consider that it is in Climate Zone 5A, which is not a warm-humid location.

1.2.2 Thermal Bridging

Ideally, every part of a building can be perfectly aligned with the insulation. The insulation meets seamlessly between wall and ceiling and around windows. But in reality, this is often not the case. Fixtures and other building elements stick out or there are difficult junctions in the structural design that make it difficult to install the insulation seamlessly. This can cause thermal bridging.

Thermal bridging occurs when a small area of floor, wall, or roof loses substantially more heat than the surrounding area. Thermal bridging can cause heat loss, condensation, moisture and mold problems, as well as expansion and contraction. Any of these effects can lead to occupant discomfort or even illness based on the severity of the effect. They can also impact the longevity of the building and building materials. That's why it is critical for contractors to be aware of potential causes and impacts of thermal bridging.

Thermal bridging can happen in any kind of building and can be an unintended consequence. It could be caused by poor installation or design. For instance, a brick wall with a structural angle supporting the brick is a huge thermal bridge. This type of structure would need to be redesigned to prevent or minimize heat loss. Thermal bridging is most commonly caused by gaps in the insulation, which lead to air leakage and heat escape. Drywall mechanics need to be especially careful when installing insulation to ensure there are no gaps and that the insulation will remain structurally sound and in position over time to prevent gaps from appearing in the future.

ASHRAE® Standard 90.1-2022 includes requirements to address the impact of thermal bridging in building envelopes. Contractors will need to calculate the value of any thermal bridges in a structure to ensure it complies with all applicable code requirements, including state or local codes.

In recent years, there has been a push to increase energy conservation in buildings. Codes are considering the operational energy of a building to ensure it is meeting energy conservation goals and has reduced energy consumption. Thermal bridging can be a major impediment to energy conservation, causing a building to be less efficient and consume more energy.

Thermal bridging: When a small area of floor, wall, or roof loses substantially more heat than the surrounding area.

Even wood or metal studs are considered thermal bridges because they create considerable breaks in the insulation. Wood and metal are also conductors that channel heat out of the building. This poses a big problem for insulation practices that have long been built on installing insulation between the wood or metal studs of a building.

Continuous installation should prevent thermal bridging (including wood and metal studs) if done correctly (*Figure 6*). The requirement to install continuous installation or an equivalent amount of insulation in some zones is one of the most substantial changes in recent years to avoid thermal bridging and reduce energy consumption. This is a major shift away from the traditional method of installing cavity insulation only.

Figure 6 Installing continuous insulation.
Source: sima/Shutterstock

1.0.0 Section Review

1. The R-value is a measure of the ability of a material to _____.
   a. resist the passage of moisture
   b. resist heat transfer
   c. allow cold air to enter a building
   d. convert water vapor into a liquid

2. You are installing insulation for a building in Broward County, Florida. In the *IECC®*, it is marked as 1A Broward*. What does that indicate?
   a. It is in Climate Zone 2 and is marine.
   b. It is in Climate Zone 2 and is dry.
   c. It is in Climate Zone 1, is marine, and is a warm-humid location.
   d. It is in Climate Zone 1, is moist, and is a warm-humid location.

2.0.0 Insulation Materials and Types

Objective

Describe insulation materials and types.

a. Define flexible insulation and explain how it is used.

b. Define loose-fill insulation and explain how it is used.

c. Define rigid or semi-rigid insulation and explain how it is used.

d. Define reflective and spray insulation and lightweight aggregates.

Performance Task

There are no Performance Tasks in this section.

Insulation materials can be divided into four general classifications (*Table 5*). These materials are used in the manufacture of five basic categories of insulation: flexible, loose-fill, rigid or semi-rigid, reflective, and miscellaneous.

These materials can generally be used to insulate both wood- and steel-framed buildings. Installers need to pay close attention to the R-values of each material to make sure the R-value of the entire wall assembly meets applicable code requirements.

TABLE 5 Insulation Materials

Classification	Material	Comments
Mineral	Rock Slag Glass fiber Vermiculite Perlite	Rock and slag are used to produce mineral wool by grinding and melting the materials and blowing them into a fine mass. It is available as loose-fill or blanket insulation.
Vegetable (natural)	Wood Sugar cane Corn stalks Cotton Cork Redwood bark Sheep's wool Straw Hemp	Many vegetable products are processed and formed into various shapes, including blankets and rigid boards. Hemp insulation is not widely used in the US.
Plastic	Polystyrene Polyurethane Polyisocyanurate Phenolic	Polyurethane can be made into polyurethane foam insulation, which is a very common type of insulation.
Metal	Foil Tin plate Copper Aluminum	Metallic insulating materials are generally applied to rigid boards or papers and used primarily for their reflective value.

The R-value of the insulation is marked on the insulation itself or its packaging (*Figure 7*).

Figure 7 Typical R-value identification.

2.1.0   Flexible Insulation

Flexible insulation: A type of insulation that is made from a flexible material.

Fiberglass: A material made of sand and recycled glass.

Batt: A flat, pre-cut piece of insulation.

Fiberglass insulation: A type of batt, roll, or loose-fill insulation that is made of extremely small pieces (or fibers) of glass.

Mineral wool insulation: A type of batt or loose-fill insulation that is made of natural stone fibers or slag.

Flexible insulation is usually manufactured from **fiberglass** in blanket form (*Figure 8*) and fiberglass or mineral wool in **batt** form (*Figure 9*). **Fiberglass insulation** is a very common form of cavity insulation and is used in most homes in the United States. **Mineral wool insulation** is becoming more popular because it provides noise insulation and fire protection, is mold resistant, and can also be used as exterior continuous insulation. In some cases, flexible insulation can also be manufactured from wood fiber or cotton and treated for resistance to fire, decay, insects, and rodents.

Blankets of flexible insulation are available in 16" or 24" widths and the batts in 15" or 23" widths. Both are furnished in thicknesses ranging from 1" to 12". The batts are packaged in flat bundles in lengths of 24", 48", or 93" and may be unfaced or faced with asphalt-laminated kraft paper or fire-resistant foil scrim kraft (FSK) with or without nailing flanges. The blankets are furnished in rolls that may be encased in asphalt-laminated kraft paper or plastic film. In most cases, they have a facing with nailing flanges. Some blankets are available with an FSK facing. Blankets and batts with kraft or film casing and/or facings are combustible and must not be left exposed in attics, walls, or floors.

Figure 9 Flexible batt insulation (mineral wool).
Source: ronstik/Shutterstock

Figure 8 Flexible blanket insulation (fiberglass).
Source: DonNichols/Getty Images

Fiberglass batt insulation at a $3\frac{1}{2}$" thickness (standard rating or high rating) may be used on exterior walls between the studs. This has been the normal insulation thickness in the past, because of the use of $3\frac{1}{2}$"-wide studs spaced 16" on center (OC). However, in the northern parts of the country, some builders have been using 2 × 6 studs spaced 16" or 24" OC, which has allowed for an increase in the wall insulation to $5\frac{1}{2}$". This thickness is ample insulation for all parts of the United States.

2.1.1   Fiberglass Insulation Safety

Flexible fiberglass is probably the first thing that comes to mind when the average person thinks of insulation. Most of us don't realize that this common material must be handled carefully. The tiny strands of glass in fiberglass insulation can irritate skin, injure eyes, and cause a variety of respiratory problems.

Fiberglass Ingredients

The primary ingredients of fiberglass are silica (sand) and recycled (previously melted) glass. The spun glass is held together with a chemical binder. The binder's ingredients include formaldehyde, phenol, and ammonia. The ammonia in the binder sometimes gives fiberglass a strong odor.

While the insulation itself does not burn, the binding material will burn off the glass fibers when the temperature rises high enough (about 350°F). For this reason, fiberglass insulation should not be used in applications that would subject the chemical binder to temperatures approaching its flashpoint.

Heat Losses

At relatively cold temperatures, a building loses heat in many ways and through different areas. These include heat lost directly through its walls, ceiling, and roof. Heat also escapes through windows, doors, gaps or cracks, and thermal bridges in the structure. That is why it is so important to make sure every part of the building is properly insulated.

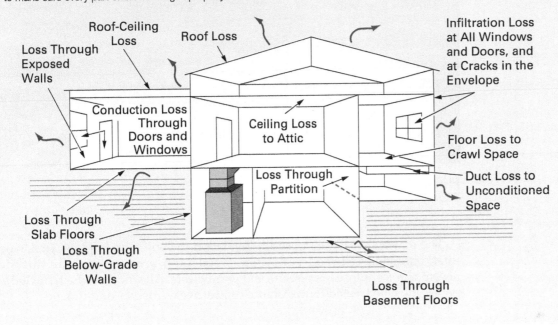

While insulation installers must wear protective equipment, their responsibilities don't end there. If debris from the installation is not properly removed or if existing insulation is disturbed, fiberglass particles could spread through the building. Fiberglass that enters an HVAC system will be carried to all parts of the building. Always use care when handling fiberglass insulation. This protects you as well as the building's current and future occupants.

2.1.2 Unfaced Versus Kraft-Faced Insulation

Unfaced insulation is fiberglass insulation that does not have a **vapor retarder**. It can be used for interior walls that do not face outside or in rooms that do not need moisture control, such as dining rooms and living rooms. It is generally noncombustible, which helps make the building more fire-resistant.

Kraft-faced insulation is fiberglass insulation that has a vapor retarder on one side. It can be installed on exterior walls and attic ceilings with the vapor retarder side facing out or in rooms that need moisture control. It generally comes in rolls or batts, which can be cut with a utility or insulation knife. The vapor retarder can be made of paper, vinyl, or foil. Make sure the type you are using complies with applicable codes. Faced insulation must also be on the outside of the insulation stack to prevent moisture from building up inside the insulation.

Unfaced insulation: Insulation that does not have a vapor retarder.

Vapor retarder: A material used to retard the flow of vapor and moisture into walls and prevent condensation within them. The vapor retarder must be located on the warm side of the wall.

Kraft-faced insulation: Insulation that has a paper, vinyl, or foil vapor retarder on one side.

2.2.0   Loose-Fill Insulation

Loose-fill insulation is a type of cavity insulation supplied in bulk form packaged in bags or bales (*Figure 10*). In new construction, it is usually blown or poured and spread over the ceiling joists in unheated attics. In existing construction that was not insulated when it was built, the material can be blown into the walls as well as the attic.

Loose-fill insulation: Insulation that comes in the form of loose material in bags or bales.

Figure 10 Loose-fill insulation.
Source: Kurteev Gennadii/Shutterstock

The materials used in loose-fill insulation include rock or glass wool, wood fiber, shredded redwood bark, cork, wood pulp products such as shredded newspaper (cellulose insulation), and vermiculite. All wood products, including paper, must be treated for resistance to fire, decay, insects, and rodents.

Heat Gains

At relatively warm temperatures, exposed walls and roofs absorb heat from the sun. Heat also enters a building through windows, doors, and gaps or cracks in the structure.

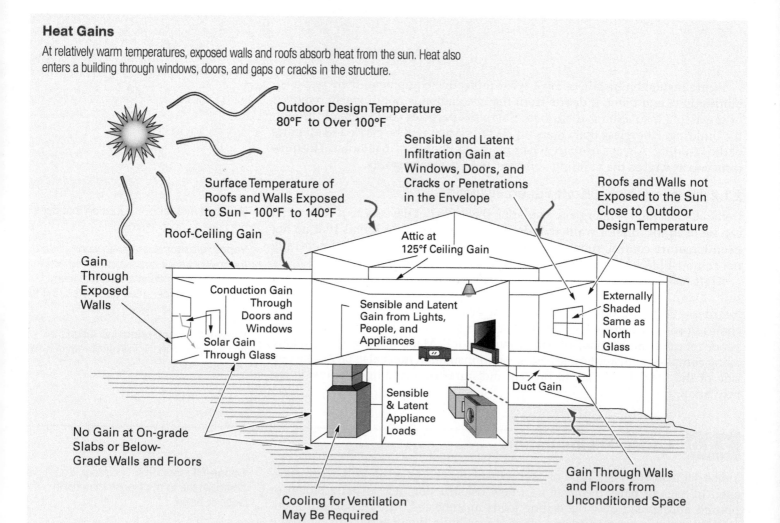

Shredded paper absorbs water easily and loses considerable R-value when damp. In addition to wall surface and/or ceiling vapor retarders, it is essential to install a waterproof membrane along the eaves to prevent water leakage.

The R-value of loose-fill insulation depends on proper application of the product. The manufacturer's instructions must be followed to obtain the correct weight per square foot of material as well as the minimum thickness. Before loose insulation is installed, the area of the space to be insulated is calculated (minus adjustments for framing members). Then, the required number of bags or pounds of insulation is determined from the bag label charts for the desired R-value.

Disadvantages of Loose-Fill Insulation

A disadvantage of loose-fill insulation is that it tends to settle over time. Insulation may have to be added to refill the cavities formed. Loose-fill insulation should not be covered by materials that could pack or crush it.

2.3.0 Rigid or Semi-Rigid Insulation

Rigid or semi-rigid insulation is available in sheet or board form and is generally divided into two groups: structural and nonstructural. It is available in widths up to 4' and lengths up to 12'.

Structural rigid foam boards come in densities ranging from 15 to 31 pounds per square foot. They are used as sheathing, roof decking, and gypsum board. Their primary purpose is structural, while their secondary purpose is insulation. The structural types are usually made of processed wood, cane, or other fibrous vegetable materials.

Nonstructural rigid foam board (*Figure 11*) or semi-rigid fiberglass insulation is usually a lightweight sheet or board made of fiberglass or foamed plastic such as polystyrene, polyurethane, polyisocyanurate, and expanded perlite. It can be used for cavity or continuous insulation. Most of these products are waterproof and can be used on the exteriors or interiors of foundations, under the perimeters of concrete slabs, over wall sheathing, and on top of roof decks. When used above grade with proper flashing, a protective and decorative coating is sometimes applied directly to the panel. This is known as an **exterior insulation finish system (EIFS)**.

In other cases, it is sealed with an air infiltration film, and normal siding is applied. The foam boards generally range in thickness from 1" to 4" with R-values up to R-30. Because all foam insulation is flammable, it cannot be left exposed. It must be covered with at least $\frac{1}{2}$" of fireproof material.

Some manufacturers also provide rigid foam cores that can be inserted in concrete blocks or used with masonry products to provide additional insulation in concrete block or masonry walls.

Rigid or semi-rigid insulation: A type of insulation that comes in formed boards made of mineral fibers.

Exterior insulation finish system (EIFS): Nonstructural, nonbearing, exterior wall cladding system that consists of an insulation board attached adhesively or mechanically, or both, to the substrate (*IBC® 2021*).

Figure 11 Rigid foam board.
Source: Whuteaster/Shutterstock

The most common rigid and semi-rigid insulation types include the following:

- *Rigid expanded polystyrene* — This material has an R-value of R-4 per inch of thickness. Water significantly reduces this value because rigid expanded polystyrene is not water-resistant. It is not recommended for below-grade insulation. It is the lowest in cost. This material is also called *beadboard*.
- *Rigid extruded polystyrene* — The R-value of this material is about R-5 per inch of thickness. It is water-resistant and can be used below grade. When used in above-grade applications, it is subject to damage by ultraviolet light and must be coated or covered.
- *Rigid polyurethane and polyisocyanurate* — Initially, these boards have R-values up to R-8 per inch. However, over time the R-value drops to between R-6 and R-7 due to escaping gases. This is referred to as *aged R-value*. These products are subject to damage by ultraviolet light and must be coated or covered.
- *Semi-rigid fiberglass* — The R-value of this material is about R-4 per inch. Boards of this kind are used on below-grade slabs and walls to provide water drainage as well as insulation. They are also used under membrane roofs as insulation. On below-grade applications, the walls or floors must be waterproofed with a coating or membrane between the wall or floor and the insulation to block water penetration.

2.4.0   Other Types of Insulation

There are other types of insulation that do not fit the previous three categories.

2.4.1 Reflective insulation

Reflective insulation: A type of insulation made of outer layers of aluminum foil bonded to inner layers of various materials.

Reflective insulation, also known as *foil-faced insulation* (*Figure 12*), usually consists of multiple outer layers of aluminum foil bonded to inner layers of various materials for strength. The number of reflecting surfaces (not the thickness of the material) determines its insulating value. Reflective insulation reflects light and can be used to protect against heat in areas that get a lot of sunlight. When used with other materials, it can warm areas. To be effective, the metal foil must face an open air space that is $3/4$" or more in depth. In some cases, reflective material is bonded to flexible insulation as the inside surface for both insulation and vapor seal purposes.

Sometimes foil is used independently of another material and is added to the other insulation. You can buy this foil insulation separately in rolls or batts. When installing it, do not push it into the insulation; it will work best with 1–2 inches of space between the foil and existing insulation. For obstacles like supports and braces, the foil can be cut, wrapped around the obstacle, and connected with foil tape. Any air or duct leaks should be sealed before installing foil insulation to avoid condensation, especially in cold climates.

> **Did You Know?**
> **Insulation Weight**
>
> Insulation materials have weight. This weight must be considered when designing a building. One of the advantages of using lightweight insulation board, such as rigid polyurethane and polyisocyanurate, is that it permits greater freedom of design. The lighter the insulation, the less weight load-bearing members of the structure must support.

> **Reflective Insulation and Heat**
>
> Reflective insulation by itself can only block radiated heat. At relatively hot temperatures, it helps keep buildings cooler by deflecting heat from the sun. At relatively cold temperatures, it can do little to prevent heat from escaping the building.

Figure 12 Reflective insulation in an attic.
Source: Ozgur Coskun/Alamy Stock Photo

2.4.2 Sprayed-in-Place Insulation

Sprayed-in-place insulation (*Figure 13*) can be applied to new or existing construction using special spray equipment. It can be injected between brick veneer and masonry walls; between open studs or joists; and inside concrete blocks, exterior wall cavities, party walls, and piping cavities. It is often left exposed for acoustical as well as insulating properties. Sprayed-in-place insulation must be applied by trained and certified contractors.

While foam is a very valuable form of insulation, drywall mechanics must assess it from a fire standpoint. Spray foam—like other types of insulation materials—is flammable. Contractors must consider the fire rating of the foam they are using when insulating a structure and ensure that the overall wall assembly fire rating is compliant with applicable codes. Foam plastics must have a thermal barrier that separates them from the building interior unless they have passed the code fire test. An acceptable barrier is $1/2"$ gypsum board.

In the 1970s, urea formaldehyde foamed-in-place insulation was injected into many homes. However, due to improper installation, the foam shrank and gave off formaldehyde fumes. As a result, its use was banned in the United States and Canada. Later, it was allowed back on the market in certain areas of the United States. A urethane foam that expands on contact can also be used. It does not have a formaldehyde problem, but it does emit cyanide gas when burned. As a result, it requires fire protection and, like urea formaldehyde, it may also be banned in some areas of the country.

Another foamed-in-place product is a phenol-based synthetic polymer called Tripolymer® that is fire-resistant and does not drip or create smoke when exposed to high heat. This material does not expand once it leaves the delivery hose of the proprietary application equipment.

Using Sprayed- and Foamed-in-Place Insulation

Sprayed-in-place insulation materials are well suited for irregular surfaces. These include walls and ceilings that are curved or that have beams, pipes, or other equipment protruding from them. Foams and sprays can be built up in layers to the desired insulation thickness.

Figure 13 Sprayed-in-place insulation.
Source: anatoliygleb/123RF

2.4.3 Lightweight Aggregates

Insulation material consisting of perlite, vermiculite, blast furnace slag, sintered clay products, or cinders is often added to concrete, concrete blocks, or plaster to improve their insulation quality and reduce heat transmission.

2.4.4 Materials for Continuous Insulation

The following types of insulation materials are generally used for continuous insulation:

- *Nonstructural rigid foam board* — One of the most common materials used for continuous insulation, rigid foam board is a firm material that is easy to install in boards over the sheathing of a building. It is also possible to fasten rigid foam board to framing or sheathing as cladding without compressing the insulation, which meets the code requirement that continuous insulation be uncompressed. Types of nonstructural rigid foam board include the following:
  - Rigid expanded polystyrene
  - Rigid extruded polystyrene
  - Foil-faced polyisocyanurate (Polyiso)
- *Mineral wool (rockwool) board* — Mineral wool is a semi-rigid type of continuous insulation material. It is often made largely out of recycled and renewable material. Because mineral wool is made of bonded rock fibers, it is noncombustible and can resist temperatures of 2,000°F, making it very fire resistant. It is vapor permeable and water resistant.
- *Spray foam* — Sprayed-in-place insulation is an effective type of continuous insulation. It can be sprayed continuously on the exterior of a building over the wood sheathing.

2.0.0 Section Review

1. Why is it important to wear protective equipment when handling fiberglass insulation?
   a. It can irritate skin, injure eyes, and cause respiratory problems.
   b. It can cause long-term heart problems.
   c. It can cause silicosis and chronic obstructive pulmonary disease (COPD).
   d. It is hot to touch and can cause burns.

2. Which type of insulation is typically blown or poured over ceiling joists in unheated attics?
   a. Flexible insulation
   b. Loose-fill insulation
   c. Semi-rigid insulation
   d. Sprayed-in-place insulation

3. Which of the following materials is likely to be used in structural rigid foam board insulation?
   a. Polystyrene
   b. Processed wood
   c. Fiberglass
   d. Expanded perlite

4. What is an acceptable thermal barrier that can be used to separate foam insulation from the building interior?
   a. Vapor retarder
   b. 2" fiberglass insulation
   c. $\frac{1}{4}$" of bonded aluminum foil
   d. $\frac{1}{2}$" gypsum board

3.0.0 Installation Guidelines

Objective

Explain how to install different types of insulation.

a. Describe how to install flexible insulation.

b. Describe how to install loose-fill insulation.

c. Describe how to install rigid or semi-rigid insulation.

d. Describe how to install foam insulation.

Performance Task

1. Install blanket insulation in a wall.

Before installation, building plans and applicable codes must be checked to determine the R-values and the types of insulation required or permitted for the structure being insulated. Then, the required amount of insulation for the structure must be calculated. Any specific instructions provided by the selected manufacturer must be followed when installing the insulation.

The estimator will estimate the amount of insulation for the walls, ceilings, and floors of a structure. Installers do not typically need to be a part of the estimation process.

3.1.0 Flexible Insulation Installation

Flexible insulation is used in unfinished floors, walls, ceilings, attics, and crawl spaces and can be fitted between joists, studs, and beams. It can insulate ducts and tank-style water heaters. Because it is not waterproof, it should not be used in foundations or other areas with high exposure to moisture.

> **WARNING!**
>
> Wear proper eye protection, respiratory equipment, and gloves when handling and installing insulation.

3.1.1 Steps for Installing Flexible Insulation

Use the following procedure when installing typical flexible insulation:

Step 1 For walls, measure the inside cavity height and add 3". From the wall, lay the distance out on the floor and mark it. Unroll blanket insulation or lay batts on the floor. Use two layers or more. At the cut mark, compress the insulation with a board and cut it with a utility knife. On blanket or faced insulation, remove about 1" of insulation from the ends to provide a stapling flange at the top and bottom.

Step 2 If a separate interior vapor seal will be installed, install blanket or faced insulation so that the stapling flange is against the inside of the wall studs (*Figure 14*).

- If the facing of the blanket or batt is the vapor seal, install the stapling flange against the inside of the stud. Do not go over the face of the stud (*Figure 15*). For faced or blanket insulation, use a power, hand, or hammer stapler to first staple the top flange to the plate.
- Align and staple down the sides.
- Staple the bottom flange to the sole plate. Pull the flanges tight and keep them flat when stapling. Follow the manufacturer's instructions for staple placement.
- For unfaced batt insulation, install the batt at the top and bottom first and push it tight against the plates.
- Evenly push the rest of the batt into the cavity (*Figure 16*).
- Spray foam insulation into the narrow spaces around windows and doors, and cover it with a plastic or tape vapor seal. Be careful to use a foam that doesn't expand so much that it distorts the window.

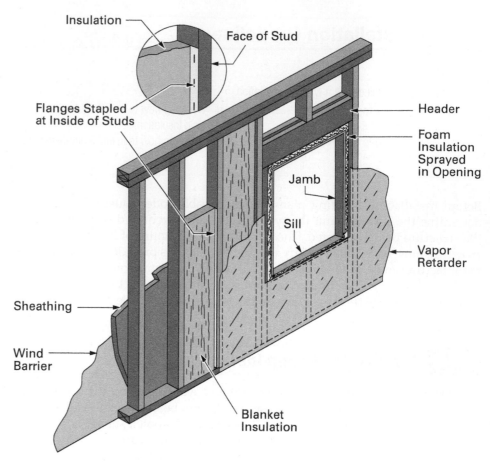

Figure 14 Blanket installation without integral vapor seal.

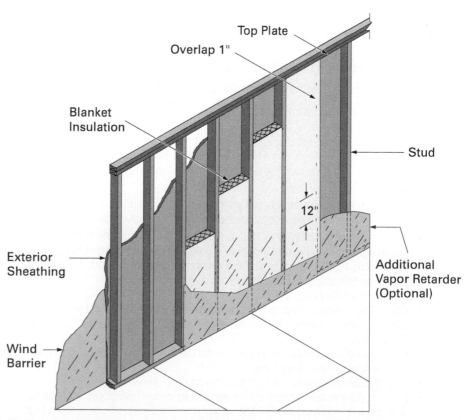

Figure 15 Blanket installation with integral vapor seal.

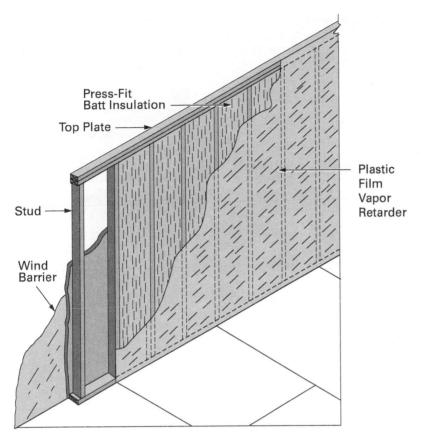

Figure 16 Batt insulation with separate vapor retarder.

Cathedral Ceilings

If a cathedral ceiling incorporates gypsum drywall attached to the bottom of the rafters, airflow must be maintained from the soffit to the ridge. Proper ventilation must also be maintained above and below skylights to prevent buildup of heat and moisture. Check your local code for the appropriate methods to use when working with cathedral ceilings.

WARNING!

Exercise caution when installing insulation around electrical outlet boxes and other wall openings or devices. Failure to do so may result in electrocution.

Step 3   Faced or blanket insulation for ceilings or floors is usually installed from the bottom in the same manner as the walls. Unfaced batts can be installed from either the top or the bottom.

- Make sure that ceiling insulation extends over the wall into the soffit area (*Figure 17*). Also make sure soffit baffles (*Figure 18*) are inserted over and cover the ceiling insulation. The baffles should be fastened to the roof deck to hold them in place so that they do not slide down into the soffit and block ventilation.

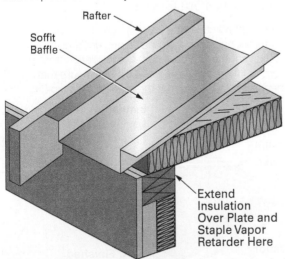

Figure 17 Ceiling insulation at wall and soffit.

Figure 18 Typical plastic soffit baffle (shown upside down).

- For floors, ensure that the insulation is installed around the perimeter of the floor against the header (*Figure 19*). Floor insulation over a basement is installed with the vapor retarder facing down.

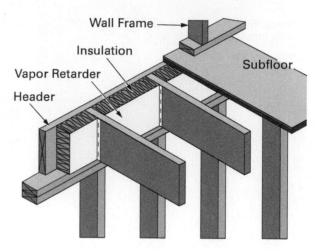

Figure 19 Perimeter floor insulation.

- Over a crawl space, the vapor retarder faces up. In either case, the insulation can be supported below by a wire mesh (chicken wire), if desired.
- Install a vapor retarder only if specified for the area. A vapor retarder on the inside face of the wall is not a good idea in certain geographical locations.

3.2.0   Loose-Fill Insulation Installation

For new construction, loose-fill insulation is used primarily for attic insulation. On older construction, it can also be blown into wall cavities through holes drilled at the center and tops of exterior walls. The following steps only cover attic or ceiling installation.

> **WARNING!**

Wear proper eye protection, respiratory equipment, and gloves when handling and installing insulation.

3.2.1 Steps for Installing Loose-Fill Insulation

Use the following procedure when installing loose-fill insulation:

Step 1   Make sure that the finished ceiling below has been installed. Also, ensure that a separate vapor retarder has been installed to prevent moisture penetration of the insulation and to prevent the fine dust from the insulation from penetrating the ceiling in the event of future cracks (*Figure 20*). Make sure that soffit baffles and blocking have been installed to prevent the material from spilling into the soffits.

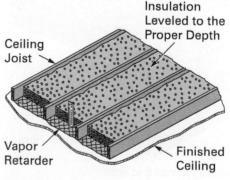

Figure 20 Loose-fill insulation.

Polystyrene Forms

Structural forms made of polystyrene are sometimes used for residential and light commercial construction. Concrete is poured into the forms, which are left in place to provide insulation for the walls. The forms usually provide sufficient insulation by themselves, but check the local code for these requirements.

Source: Oleksandr Rado/Alamy Stock Photo

Step 2 Add markers throughout the area before pouring in the loose-fill insulation to indicate the initial installed thickness. This is a code requirement that allows an inspector to easily see the depth of the insulation. There should be one marker per 300 square feet. Markers can be attached to trusses or joists and should be marked with numbers that are at least 1 inch in height for visibility. The markers must face the access opening of the attic.

Step 3 Pour the insulation from bags or blow the insulation over the ceiling joists using special equipment. Using a straightedge, tamp the insulation and then level it to the required depth for the R-value desired (*Figure 21*).

Figure 21 Leveling loose-fill insulation.
Source: Dorling Kindersley ltd/Alamy Stock Photo

3.3.0 Rigid Insulation Installation

Rigid insulation panels can be fastened like sheathing over the studs or wood sheathing of a structure as exterior continuous insulation. Nails with large heads or washers or screws with washers are used to prevent crushing or compressing the insulation (*Figure 22*).

This type of insulation is particularly important for steel-framed buildings to counteract the natural heat conduction of steel. When insulation is installed outside of a steel frame, it prevents heat from escaping through the steel. When cavity insulation is installed between steel framing members, it is often not sufficient to prevent heat from escaping through the steel studs.

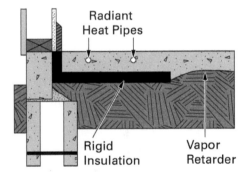

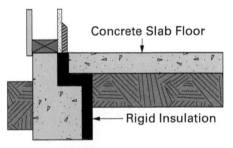

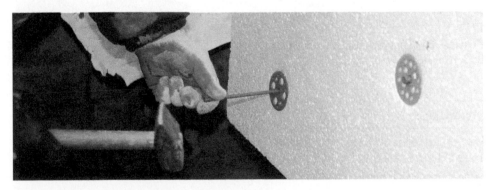

Figure 22 Installing rigid board insulation.
Source: Radovan1/Shutterstock

Rigid insulation panels may be installed on the exterior of a foundation. Typically, the exterior of the foundation is waterproofed first. Then, the panels are applied over special mastic and secured with concrete nails to hold them in place until the mastic sets. For existing construction, the panels may be installed on the interior of the foundation if the walls are adequately waterproofed.

Figure 23 shows typical methods of installing rigid insulation under surface slabs. Slabs lose energy when heat is conducted out and through the slab perimeter. Usually, insulation is only applied around the perimeter of the slab, which helps prevent this heat loss. It can be installed downward from the top of the

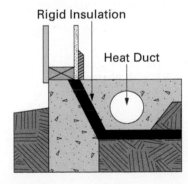

Figure 23 Rigid insulation installed under a concrete slab.

slab on the inside or outside of the foundation wall. If a slab is separate from a foundation wall, insulation can be installed between the wall and the slab or on the exterior of the wall. A vapor retarder should be applied under the slab and over any insulation under the slab.

Insulation below grade is required in Climate Zones 4–8. It should extend vertically, under the slab, and perpendicularly by the amount indicated in Table N1102.1.2 in Section R402 of the *IECC*®. Insulation that extends away from the building must be protected by pavement or at least 10 inches of soil.

3.4.0 | Foam Insulation Installation

Foam is sprayed between framing members rather than placed and attached like flexible and rigid insulation (see *Figure 24*). It is particularly useful for insulating small, tough-to-access spaces, such as the cavities around doors and windows. Spray foam is messy and should be around a certain temperature, so it is important to prepare fully before beginning the installation.

Figure 24 Sprayed-in-place insulation.
Source: anatoliy_gleb/Shutterstock

WARNING!

Remember that both foamed-in-place and sprayed-in-place insulation should be installed only by trained contractors.

3.4.1 Steps for Installing Foam Insulation

Installers should follow the instructions for using the certain foam insulation systems provided by the manufacturer. The following steps are general guidelines that apply to most foam insulation installation processes:

Step 1   Make sure the foam canisters are between 70 and 85 degrees, although this may vary by manufacturer. Check the manufacturer's specifications for the best results. Some manufacturers may also recommend a certain temperature for the surface on which you will be spraying the foam.

Step 2   Prepare the area. Use plastic to cover everything that shouldn't be touched by foam. Make sure to cover outlets and loose cords as well.

Step 3   Wear the appropriate PPE. It is extremely important to avoid all foam inhalation and skin contact. PPE includes the following:

- Disposable suit that covers the head
- Air respirator mask
- Eye protection
- Disposable gloves
- Shoe coverings

Step 4   Once you've prepared, you can begin spraying the foam. The R-value you need to meet will indicate how many inches of foam should be applied. Remember that the foam will expand, so it's important to pay attention to the type of foam and how much it will expand over time. If you are combining foam insulation with another insulation material, you may only need a single layer of foam.

Step 5   Use a knife to remove any extra foam over the framing members.

Foam insulation can be applied to the exterior sheathing of a structure to create continuous insulation. When applied on the exterior, it also acts as an air and water-resistive barrier. It may be more effective in this use because when applied in cavities, it is still being interrupted by thermal bridging (the studs). On the exterior of a structure, it can be applied continuously to create an uninterrupted layer of insulation.

3.0.0 Section Review

1. When installing faced insulation in a wall that will have a separate vapor barrier installed, the staples should be fastened to the wall frame _____.
   a. on the faces of the wall studs, top plate, and sole plate
   b. on the inside surfaces of the wall studs, top plate, and sole plate
   c. only to the top plate and sole plate
   d. only to the faces of the wall studs

2. How many markers should be placed in loose-fill insulation per 300 square feet?
   a. 1
   b. 2
   c. 3
   d. 4

3. What is a type of fastener you can use when installing rigid insulation panels to prevent crushing the insulation?
   a. Screws with washers
   b. Nails with small heads
   c. Sheathing tape
   d. Wood glue

4. What should you do before installing foam insulation?
   a. Waterproof the exterior of the foundation.
   b. Install soffit baffles and blocking.
   c. Use plastic to cover everything that shouldn't be touched by foam.
   d. Permanently install strike-off boards.

4.0.0 Moisture Control

Performance Task	**Objective**
2. Install a vapor retarder on a wall.	Explain how moisture control is accomplished in the construction of a building. a. Explain how to design a structure to control moisture. b. Describe how ventilation and vapor retarders are used for moisture control and explain how they are installed.

Water vapor: Water in a vapor (gas) form, especially when below the boiling point and diffused in the atmosphere.

Water vapor contained in air can readily pass through most building materials used for wall construction. This vapor caused no problem when walls were porous because it could pass from the warm wall to the outside of the building before it could condense into liquid water (*Figure 25*).

When buildings were first constructed with insulation in the walls to cut down on heat loss, moisture in the air passed through the insulation until it reached a point cold enough to cause it to condense. The condensed moisture froze in very cold weather and reduced the efficiency of the insulation. The ice contained within the wall thawed as the weather warmed, and the resulting water in the wall caused studs and sills to decay over time.

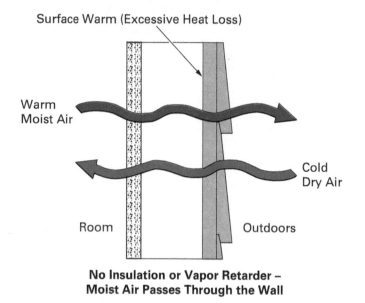

**No Insulation or Vapor Retarder –
Moist Air Passes Through the Wall**

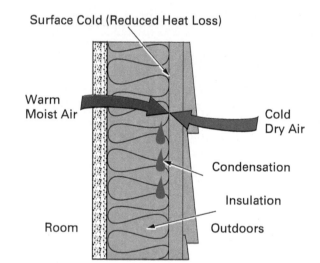

**Insulation Only – Moist Air Passes Into Wall
and Condenses, Causing Damage
to Insulation and Building Framework**

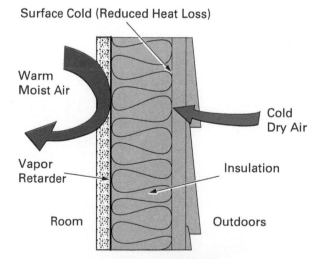

**Insulation with Vapor Retarder –
Moisture Cannot Enter Wall Space**

Figure 25 Effects of insulation and vapor retarder.

For these reasons, it is important to keep cellars, basements, crawl spaces, exterior walls, and attics dry. Moisture in crawl spaces, basements, and attics also encourages wood-chewing insects such as termites, as well as the growth of mold. In the case of crawl spaces, moisture often rises from the ground into the crawl space during periods of heavy rain.

4.1.0 Designing a Structure to Control Moisture

To prevent the concentration of this damaging moisture, some precautions must be taken in the original design of the structure:

- The earth must slope down and away about 20' from the structure, carrying surface water away.
- The crawl space should be protected from moisture by a vapor retarder on the ground.
- The foundation walls should be penetrated with vents so that moisture will not be trapped in the crawl space.
- A vapor retarder should be installed between the insulation and the subfloor.

4.1.1 Moisture Control in Basements

Basements usually have the most trouble with **condensation** in summer during humid weather. The earth under the concrete basement floor is comparatively cool, causing the floor of the basement to be a cold surface. The hot air is saturated with moisture and condenses when it comes in contact with the cooler surfaces of the floor and walls. This problem is difficult to control. If the surface of the concrete is rough and porous, the moisture will sink in and not cause a wetness problem. If, however, the floor is dense and smoothly finished, the tightly knit grains of concrete form a vapor retarder of sorts, and the water collects on the slab. This problem can usually be solved by using dehumidification devices during the summer months.

Condensation: The process by which a vapor is converted to a liquid, such as the conversion of the moisture in air to water.

Moisture seeping through the concrete floor is a different problem. In new construction, this is controlled by installing perimeter drainage and a vapor retarder under the concrete slab. When installing polyethylene film as an underslab vapor retarder, be careful not to tear, puncture, or damage the film in any way. Any passageways for moisture will defeat the purpose of the vapor retarder. Prior to pouring the concrete slab, make sure the polyethylene film is placed properly and is free of punctures. Keep all construction debris away from the vapor retarder.

To keep moisture from rising up into the basement, 6" of coarse gravel should be placed over the compacted earth to provide drainage to the perimeter drain before the slab is poured. A polyethylene film should be placed on top of the gravel to keep the concrete from penetrating into the gravel and possibly weakening the slab. In very wet areas or areas with a high water table, floor drainage, in addition to a gravel bed, may also be required.

4.1.2 Continuous Insulation

Continuous insulation helps to control moisture in several ways. It seals off any gaps in the wall assembly, which prevents vapor from entering. It creates an air barrier that makes the structure airtight and helps reduce air leakage, making it easier to control air flow and, by extension, vapor flow. Continuous insulation should also provide an effective thermal barrier that regulates the temperature of the wall. When the air is hot on the outside of a structure, this should prevent condensation from building up inside the structure.

4.2.0 Interior Ventilation

One of the best ways to reduce or eliminate the chances of moisture damage in attics or in the space between the rafters and the finished roof is through proper ventilation. Ventilation provides a stream of outside air to remove trapped moisture before it is allowed to do any damage. In insulated attics, baffles (blocking strips) are used to keep the insulation material from getting into the vented areas. With the increased use of blown-in insulation in attics, baffles are being required by code in some areas.

The amount of ventilation required varies by climate and building codes. Attics and gable and hip roofs may be ventilated with a variety of louvers and vents. Flat roofs are ventilated with a combination of eave vents and roof stacks (*Figure 26*).

Ice dams can usually be avoided by installing plenty of insulation and providing ample ventilation in the attic.

Properly designed subroof ventilation is the best weapon for preventing water vapor infiltration into a steeply sloped roof, but it is less effective on roofs with low slopes because natural **convection** decreases with diminishing roof height. Moisture dissipation occurs through **diffusion** and wind-induced ventilation.

Normally, the ventilation requirement for a gable roof is 1 ft^2 of free air ventilation for every 300 ft^2 of ceiling area if a vapor retarder exists under the ceiling. If no vapor retarder is present, the requirement is 1 ft^2 for every 150 ft^2 of ceiling area. The total requirement must be split evenly between the inlet vents and the outlet vents.

Free air ventilation is the rating of the ventilation devices, taking into account any restrictions caused by screening, louvers, and other devices.

Convection: The movement of heat that either occurs naturally due to temperature differences or is forced by a fan or pump.

Diffusion: The movement, often contrary to gravity, of molecules of gas in all directions, causing them to intermingle.

Did You Know?

Ice Dams

In colder climates, ice dams can be a problem. Ice dams are formed along the edge of a sloping roof when a building's attic is not properly insulated and ventilated. Heat escaping through the roof melts accumulated snow, forming icicles along the edge of the roof. Over time, water collects under the outer layer of snow and is trapped by the ice. This water backs up under the shingles and penetrates the roof, causing water damage and other problems.

In areas with a history of water damage to structures from ice dams at roof eaves, an ice barrier is required by code for added protection. The ice barrier consists of self-adhering polymer-modified bitumen material or two layers of cemented underlayment, which must extend from the eave to at least 24" inside the exterior wall line of the building.

Source: LegART/Getty Images

Did You Know?
Mold

Moisture accumulating inside a building can damage the structure and promote the growth of mold. While it is not always harmful, this mold may cause allergic reactions or other respiratory problems in some people. Airborne mold spores can also cause infections, primarily in people whose immune systems are compromised.

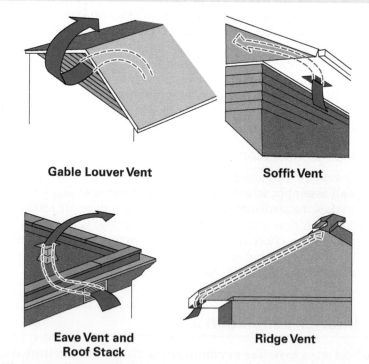

Gable Louver Vent   **Soffit Vent**

Eave Vent and Roof Stack   **Ridge Vent**

Figure 26 Various methods of roof ventilation.

4.2.1 Vapor Retarders

Vapor retarders are an important part of moisture control. A vapor retarder, also known as a *vapor diffusion retarder (VDR)*, is any material or substance that will not permit the passage of water vapor or will do so only at an extremely slow rate. When vapor reaches its **dew point**, which can occur inside the insulation or at the cool outer surface, it condenses. This results in a reduction or total loss of the thermal efficiency of the insulation, as well as dripping and damage. To prevent this process, select a vapor retarder that is both easy to apply and resistant to jobsite abuse as well as compliant with any code requirements. A properly installed vapor retarder will protect ceilings, walls, and floors from moisture originating within a heated space.

Different types of vapor retarders are required in certain climate zones (see *Table 6*). It is important to check all applicable codes to ensure you are installing the right type of vapor retarder. The *IRC*® provides a rating system that separates vapor retarders into different classes based on their **permeance**. This helps clarify which types of vapor retarders are best used in which climate zones. There are three classes, which are known as the material's **vapor retarder class** (see *Table 7*).

Vapor permeance refers to how a material allows water vapor to pass through it. A **vapor permeable** material permits the passage of moisture vapor and has a moisture vapor permeance of 5 perms or greater, as defined by the *IBC*®.

Dew point: The temperature at which air becomes oversaturated with moisture and the moisture condenses.

Permeance: The ratio of water vapor flow to the vapor pressure difference between two surfaces.

Vapor retarder class: A measure of a material's ability to limit the amount of moisture that passes through it.

Vapor permeable: Permitting the passage of moisture vapor. A vapor permeable material, as defined by the *IBC*®, has a moisture vapor permeance of 5 perms or greater.

TABLE 6 IRC® Vapor Retarder Requirements by Zone

Climate Zone	Marine 4	5	6	7	8
Class I or II[1] vapor retarders required on the interior side of frame walls[2]	Yes	Yes	Yes	Yes	Yes
Class III[1] vapor retarder permitted for:					
Vented cladding over wood structural panels	Yes	Yes			
Vented cladding over fiberboard	Yes	Yes	Yes		
Vented cladding over gypsum	Yes	Yes	Yes		
Continuous insulation with R-value greater or equal to 2.5 over 2 × 4 wall	Yes				
Continuous insulation with R-value greater or equal to 5 over 2 × 4 wall		Yes			
Continuous insulation with R-value greater or equal to 7.5 over 2 × 4 wall			Yes		
Continuous insulation with R-value greater or equal to 10 over 2 × 4 wall				Yes	Yes
Continuous insulation with R-value greater or equal to 3.75 over 2 × 6 wall	Yes				
Continuous insulation with R-value greater or equal to 7.5 over 2 × 6 wall		Yes			
Continuous insulation with R-value greater or equal to 11.25 over 2 × 6 wall			Yes		
Continuous insulation with R-value greater or equal to 15 over 2 × 6 wall				Yes	Yes

[1]Class I: sheet polyethylene, unperforated aluminum foil; Class II: kraft-faced fiberglass batts; Class III: latex or enamel paint
[2]Exception for basement walls, the below-grade portion of any wall, and construction where moisture or freezing will not damage any materials

Some vapor retarder materials, such as kraft paper, are attached to blanket or batt insulation. They are installed when insulation is installed. Others, such as aluminum foil, may be applied to the back of gypsum board during its installation.

TABLE 7 Vapor Retarder Classes

Class	Definition	Examples
I	0.1 perms or less	Sheet polyethylene Nonperforated aluminum foil
II	Greater than 0.1 to less than or equal to 1.0 perms	Kraft-faced fiberglass batts Vapor retarder paint
III	Greater than 1.0 perms to less than or equal to 10 perms	Latex paint Enamel paint

2021 IRC Table R702.7(1) provides a list of materials that can be used for each class of vapor retarder and are deemed to comply with the test standard. No testing is required for these materials. All other materials are required to be tested.

Permeability: The measure of a material's capacity to allow the passage of liquids or gases.

Perm: The measure of water vapor permeability. It equals the number of grains of water vapor passing through a 1 ft² piece of material per hour, per inch of mercury difference in vapor pressure.

Polyethylene film used as a vapor retarder is applied over studs and ceiling joists after insulation is installed. Vapor retarders are also installed under slabs, between the gravel cushion and the poured concrete.

The **permeability** of a substance is a measure of its capacity to allow the passage of liquids or gases. Water vapor permeability is the property of a substance to permit the passage of water vapor and is equal to the permeance of a substance that is 1" thick.

The measure of water vapor permeability is called **perm**. This equals the number of grains of water vapor passing through a 1 ft² piece of material per hour, per inch of mercury difference in vapor pressure. All you need to remember is that any material that has a perm rating of 1.0 or less is considered a vapor retarder and will not allow the passage of any appreciable or harmful amounts of water vapor. Any material with a rating higher than 1.0 is a breathable material that will permit the passage of water vapor in whatever degree its perm rating indicates. The higher the perm number, the greater the amount of water vapor that will pass through the material in a given time; 0.0 is totally impermeable.

A properly installed vapor retarder will protect ceilings, walls, and floors from moisture originating within a heated space (*Figure 27*). Note that the air flow in this image does not reflect all climates. This will vary depending on the climate zone.

Smart Vapor Retarders

Naturally changing weather conditions present problems when installing vapor retarders. It is important to have a vapor retarder that is doing the right thing at the right time, depending on the conditions. In the past, it was advised to install a vapor retarder on the "warm side" of the wall, which meant the inside of the wall in colder climates or the outside of the wall in warmer climates. But this method can get confusing in places where sometimes it's warmer outside while other times of the year, it's warmer inside.

One solution is to use a smart vapor retarder. These vapor retarders have a permeance that varies based on humidity. These can achieve high permeance in summer to adapt to the higher humidity and low permeance in winter for low humidity, preventing condensation and moisture.

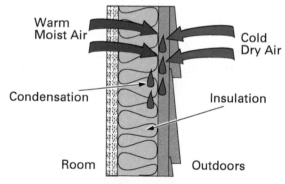

Wall with No Vapor Retarder

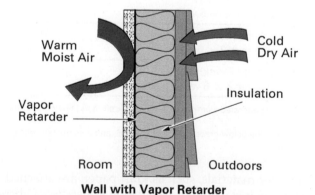

Wall with Vapor Retarder

Figure 27 Vapor retarder installation.

An insulated wall will divide two temperature gradients. The area on the inside of the structure will normally be warmer than the air on the outside. The vapor retarder is usually located on the warm side to prevent moisture from moving through the insulation to the cool side and condensing.

Materials

Common vapor retarder materials include asphalted kraft paper, aluminum foil, and polyethylene film.

Asphalted kraft paper is usually incorporated with blanket or batt insulation. It serves as a means for attaching the insulation to the building framework and as a reasonably good vapor retarder when installed on the warm side of the wall or ceiling (*Figure 28*).

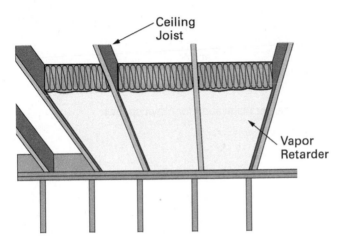

Figure 28 Installing insulation batts between ceiling joists with vapor retarder down.

Aluminum foil may be incorporated with blanket or batt insulation in the same manner as kraft paper. It is also applied to the back of gypsum lath and gypsum board where it works as a relatively effective vapor retarder.

Polyethylene film is applied over the studs and ceiling joists after the insulation is installed. When gypsum board with polyethylene film or foil backing is used, the insulation will normally be plain batts or blankets that do not have an integral vapor retarder. As a vapor retarder, polyethylene film is stapled over the studs and also covers the window frames. This helps to keep the window frames and sashes clean during application and finishing of the gypsum board. The film should be overlapped 2" to 4" and sealed with special mastic or tape.

4.2.2 Installing Vapor Retarders

There are different ways to install vapor retarders depending on the part of the structure.

Installation in Crawl Spaces

The exposed ground under an unventilated crawl space should be covered with a continuous Class I vapor retarder ground cover to protect the underside of the house from condensation (see *Figure 29*). Each joint of the vapor retarder should be taped or sealed and overlap by 6". The edges of the vapor retarder should be attached to the stem wall and extend at least 6" up the wall.

Besides installing vapor retarders, crawl spaces should be properly vented to permit the escape of moisture. Usually this is accomplished by installing a proper number of screened foundation vents in the above-grade foundation surrounding the crawl space. The normal requirement is 1 ft^2 of free air ventilation for every 150 ft^2 of crawl space area when a vapor retarder ground cover is used.

Vapor Retarder Backing

Many of the insulation materials produced today have a vapor retarder applied to the inside surface. Many interior wall surface materials are also backed with vapor retarders. When these materials are properly installed, they usually provide satisfactory resistance to moisture penetration. If the insulating materials do not include a vapor retarder, then one should be installed as a separate element.

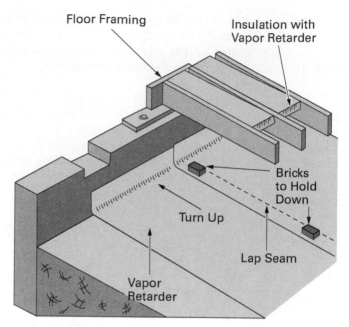

Figure 29 Vapor retarder installation for crawl spaces.

Installation in Slabs

When allowed to proceed unchecked, moisture will migrate from the ground upward through concrete and into the building, where it can cause moisture problems, damage, and higher energy costs. Even though the water table may be several feet below the slab, moisture vapor will migrate up to and through concrete slabs. Up to 80 percent of the moisture entering a structure does so by migrating from the ground beneath the structure. Moisture vapor passes through concrete more readily than liquid moisture.

Moisture in a building can cause deterioration of interior finishes, especially floors and equipment. Moisture can also add to energy costs by raising humidity and taxing cooling systems that require dehumidification.

Place a 6-mil (0.006 inch) approved vapor retarder between the base course and prepared subgrade. (Some codes may require a 10-mil vapor retarder.) The base course is sand, gravel, or crushed stone, concrete, or blast-furnace slag in a sieve on subgrade where the slab is below grade. Vapor retarders should be continuous under the slab. Take great care not to tear or puncture the retarder. Keep all construction debris away from the retarder location. Vapor retarder installation must be done by qualified contractors. When used in thickened-edge slab construction, as shown in *Figure 30*, a vapor retarder is placed between the gravel cushion and the poured concrete. The same arrangement is used for other types of slab-on-grade construction.

Figure 31 shows a method of constructing a finished floor over a concrete slab, which affords double protection against moisture. The sealer or waterproofer is placed on the slab itself, and a vapor retarder is suspended above the slab.

There are some exceptions. Vapor retarders are not required by the *IECC*® for the following:

- Any unheated accessory structures, such as garages and utility buildings
- Carports and unheated storage rooms of less than 70 ft²
- Flatwork that is unenclosed and unheated, such as driveways, walks, and patios
- Areas that have been approved by the building official as not needing a vapor retarder

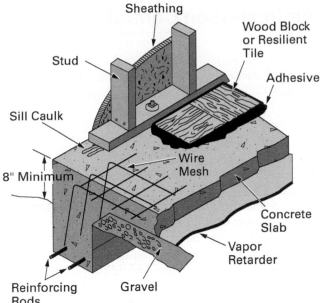

Figure 30 Thickened-edge slab vapor retarder installation.

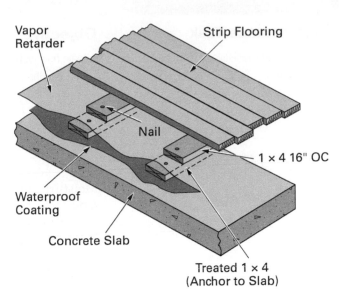

Figure 31 Surface-mounted vapor retarder on a slab.

Installation in Walls

A polyethylene sheet vapor retarder is easy to apply to frame walls where no integral barrier is provided or where a supplementary barrier is preferred. A flap would normally overlap both floor and ceiling barriers to seal the interior off completely. Adjacent sheets of the film are overlapped 2" to 4" and are sealed with a special mastic or tape.

When vapor retarders are applied to walls, particular attention should be paid to fitting the material around electrical outlet boxes, exhaust fans, light fixtures, registers, and plumbing. Considerable water vapor can escape through the cracks around the equipment, travel from the warm side of the wall to the cold side, and condense on the sheathing or siding. This is especially true if the insulation is poorly fitted at the top and bottom.

4.0.0 Section Review

1. The area of a house that typically has the most trouble with condensation during the summer is the _____.
   a. kitchen
   b. attic
   c. basement
   d. bathroom

2. The best way to prevent moisture damage in an attic is through _____.
   a. drainage channels on the roof deck
   b. moisture-absorbing materials in the insulation
   c. automated electric air dryers
   d. ventilation of the attic space

5.0.0 Water and Air Infiltration Control

Performance Task

3. Install water-resistive barriers and materials.

Objective

Explain how water and air infiltration control are accomplished in the construction of a building.

a. Describe how water-resistive barriers are used to resist liquid water.
b. Describe how air barriers are used and installed.
c. Explain how to protect openings, penetrations, and joints.

Moisture can enter a building in numerous ways. It can penetrate the interior of a building as liquid water. Rain or snow will introduce water to the exterior of a building throughout the year in varying amounts depending on the local climate. Some places may have different waterproofing requirements depending on higher amounts of rain or snow. No matter where a building is, it is important to protect the interior from liquid water.

Moisture can also be carried into a building in air that enters from the outside. When air leaks into a building, it can cause changes in temperature that create condensation on building surfaces.

Installers must take additional steps to control moisture by waterproofing a building, integrating water-resistive barriers, and preventing air infiltration.

5.1.0 Water-Resistive Barriers

Flashing: Thin, water-resistant material that prevents water seepage into a building and directs the flow of moisture in walls.

Water-resistive barrier: One or more materials installed behind exterior wall coverings to prevent water from entering a building.

The *IBC*® requires that exterior walls provide the building with a weather-resistant exterior wall envelope. **Flashing** and water-resistive barriers are essential components of this envelope. A water-resistant exterior wall envelope is not required over concrete or masonry walls.

A **water-resistive barrier** (WRB) prevents the passage of water into a building. It works together with vapor barriers to protect the building from moisture. WRB materials are installed behind exterior wall coverings to the studs or sheathing. When water does penetrate the exterior, the WRB should be ready to resist it from intruding deeper into the wall. The materials should be installed continuously over expansion and control joints.

WRBs can be made of sheet materials, fluid-applied materials, and some types of insulating sheathing. Follow any manufacturer instructions when installing them. After applying WRBs in sheets, you should check your work for any holes or tears.

Various types of WRBs include the following:

- *Asphalt felt* — This material (*Figure 32*) is generally made from recycled paper products, such as corrugated paper and sawdust, which are coated with asphalt. It has a high permeance when wet and is able to soak up water—rather than repelling it—and let it dry.

- *Asphalt-saturated grade D paper* — Similar to asphalt felt in composition, this WRB is made of asphalt impregnated into kraft paper. Is generally less expensive than asphalt felt but tends to rot if it stays wet. It comes in large rolls that should be applied over gypsum sheathing using 2-ply application. The time it can spend exposed is very limited and it needs to be covered as soon as possible. It should not be used below grade, horizontally, or for roofs.

- *Rigid foam plastic insulating sheathing* — Expanded or extruded polyisocyanurate or polystyrene panels can be applied over exterior framing. The material must have been tested as a water barrier. Joints must be carefully taped and sealed.

- *Polyolefin water barriers* — This WRB is made of spun-bonded polyolefin and should be applied over sheathing. It can be used on vertical surfaces only and should not be used below grade.

Figure 32 Asphalt roofing felt.
Source: Oleksandr Rado/Alamy Stock Photo

- *Self-adhered water-resistive membrane* — While most paper WRBs are applied mechanically, self-adhered WRBs are applied with adhesive. This material is made of rubberized asphalt compound that has been laminated to an engineered film. It comes in large rolls and should be applied over sheathing. It can only be applied when the air and surface temperature is above 40°F and should be covered as soon as possible.

- *Fluid-applied* — This is an acrylic-based material that, when sprayed or rolled across a structure, forms a seamless membrane. It can also double as an air barrier. Fluid-applied WRB has specific temperature requirements and should be applied only when the surface and air temperature is between 40°F and 100°F. It must be protected with cladding within 6 months. It should not be used below grade or for surfaces with standing water and should not be used to span joints in sheathing that are over $\frac{1}{8}$" wide.

In general, WRBs do not also act as air barriers; however, some materials can be used as both an air barrier and WRB. Liquid-applied WRBs and Zip System® sheathing can act as both types of barriers. Any WRB must be tested as a functioning water-resistive system to meet code requirements.

5.2.0   Air Infiltration Control

In addition to insulation, the exterior sheathing of a structure should be covered to prevent wind pressure from causing infiltration of outside air into the structure. To achieve maximum energy efficiency in a structure, air infiltration must be strictly controlled. Air leakage can also be prevented by installing air barriers in the interior of a structure.

5.2.1 Air Barriers

An **air barrier** in the wall assembly protects the structure from air leaks.

When there are gaps in a wall assembly, air can leak out of or into a structure. This can cause heat loss and temperature changes, which can lead to condensation and moisture damage. It can allow moisture in the air to enter a structure from outside. It can also allow rodents and insects to enter the structure. To keep the structure airtight, multiple air barriers can work together to seal the building. When a structure's air barrier is working effectively, it is reducing heat loss and energy consumption while avoiding any potential moisture damage that could result from air leaks.

An air barrier is one or more materials joined together continuously to prevent or restrict the passage of air through a building's thermal envelope and assemblies. It is often a good idea to have multiple air barriers to provide as much protection as possible. The *IRC*® requires that a continuous air barrier be installed in the building envelope and exterior thermal envelope. Any joints or breaks in the air barrier must be sealed. Drywall mechanics should ensure the insulation aligns with the air barrier they are using for any space.

Air barriers should be:

- Continuous (no gaps)
- Impermeable to air leakage
- Durable (should last as long as the building)
- Repairable
- Rigid (supported from wind pressure)

Air barriers are required by code unless the structure has one of these systems:

- Concrete masonry walls coated with block filler and two applications of a sealer or paint coating
- Portland cement/sand parge, plaster, or stucco that is at least $\frac{1}{2}$" thick

Air barrier: One or more materials joined together continuously to prevent or restrict the passage of air through a building's thermal envelope and assemblies.

5.2.2 Interior Air Barriers

Gypsum board can serve as an effective interior air barrier when sealed continuously throughout a building. It must be sealed in the following locations:

- Around the rough openings of doors and windows
- To the first stud in the wall
- Along the top and bottom plates on exterior walls
- At the top plate of partitions when adjacent to an unconditioned space

Spray foam insulation can also be used as an air barrier, as can other forms of air impermeable insulation.

Air barriers generally need to meet certain air permeance requirements that may vary by region. Make sure to check all applicable codes when installing air barriers.

5.2.3 Exterior Air Barriers

Different types of materials are used to protect different types of residential or commercial buildings. Some materials can act as both air barriers and WRBs.

Wraps

House wraps or building wraps are most commonly used to prevent water and air leakage in residential buildings (*Figure 33*). These products are sold under brand names such as Tyvek® or ProWrap®. When properly applied and sealed, these wraps provide a nearly airtight structure no matter what sheathing material is used. Most versions of these wraps are an excellent secondary barrier under all siding, including stucco and EIFS.

Figure 33 Residential building wrap.
Source: B Christopher/Alamy Stock Photo

Building wrap can also be applied to commercial buildings (*Figure 34*).

Nails with large heads, nails or screws with plastic washers (*Figure 35A*), or 1" wide staples may be used to secure the wrap to wood, plastic, insulating board, or exterior gypsum board. Screws and washers are used for steel construction. Special contractor's tape (*Figure 35B*) or sealants compatible with the wrap are used to seal the edges and joints of the wrap.

WARNING!

Some building wraps are slippery and should not be used in any application where they can be walked on. Because the surface will be slippery, use pump jacks or scaffolding for exterior work above the lower floor. Take extra precautions when using ladders to prevent sliding on the wrap.

(A) Nails With Plastic Washers

(B) Contractor's Tape

Figure 34 Commercial building wrap.
Source: Peter Titmuss/Alamy Stock Photo

Figure 35 Building wrap accessories.
Sources: (A) Jonny White/Alamy Stock Photo; (B) Alysson M/Shutterstock

Always refer to the manufacturer's instructions for specific installation information. House or building wrap is generally installed as follows:

Step 1 Using two people and beginning at a corner on one side of the structure, leave 6" to 12" of the wrap extended beyond the corner to be used as an overlap on the adjacent side of the structure. Align the roll vertically and unroll it for a short distance. Check that the stud marks on the wrap align with the studs of the structure. Also check that the bottom edge of the wrap extends over and runs along the line of the foundation. Secure the wrap to the corner at 12" to 18" intervals.

Step 2 Continue around the structure, covering all openings. If a new roll is started, overlap the end of the previous roll 6" to 12" to align the stud marks of the new roll with the studs of the structure. Repeat these steps for the upper parts of the structure, making sure the top and bottom layers of wrap overlap by 6" to 12".

Step 3 At the top plate, make sure the wrap covers both the lower and upper (double) top plate, but leave the flap loose for the time being.

Step 4 At each opening, cut back the wrap using one of these methods:
Method 1: Fold the three flaps around the sides and bottom of the opening and secure every 6". Trim off the excess.
- At the outside, install 6" flashing along the bottom of the opening, then up the sides over the top of the wrap. Install head flashing at the top of the opening under the wrap and over the side flashing. Tape the flap ends to the head flashing.

Figure 37 Worker pointing to an expansion joint in concrete.
Source: Sumith Nunkham/Getty Images

Section R703 of the *IRC*® indicates what type of joint treatment should be applied for different types of siding material. In general, WRBs should be applied horizontally and overlap at least 6" where joints occur. Horizontal joints in panel siding must be lapped by at least 1" or shiplapped or flashed with Z-flashing over wood panel sheathing. Joints can be sealed with the following materials:

- *Caulk* — Caulk is generally better suited for filling smaller gaps and is typically applied from a tube.
- *Tape* — Approved types of joint bridging tapes can be very effective for filling joints. Strips of tape should be cut to the appropriate width.
- *Foam seals* — Foam seals should generally be used for joint widths with a maximum of 8". Open-cell foams are better suited for vertical applications because they allow flow-through of moisture. Closed-cell foams are watertight and are better suited for horizontal applications.

The sealant must be able to insulate the joint while allowing it to move over time up to its full range of movement. That means it should be flexible. Some sealants may work better for certain climates. Evaluate the project needs and code requirements before selecting and applying a sealant. Follow specific manufacturer requirements to fill and seal joints.

5.0.0 Section Review

1. Which type of water-resistive barrier should be applied only when the surface and air temperature is between 40°F and 100°F?
   a. Asphalt felt
   b. Foam plastic insulating sheathing
   c. Polyolefin water barriers
   d. Fluid-applied

2. Which of the following can serve as an effective interior air barrier?
   a. Gypsum board
   b. Building wraps
   c. Liquid applied
   d. Rigid foam sheathing

3. What is flashing?
   a. A type of joint that allows a building to move when temperature changes cause any type of structural movement
   b. A flat, pre-cut piece of insulation that is often made of fiberglass or mineral wool
   c. A sturdy, durable type of continuous air barrier made of plywood and OSB boards
   d. Thin, water-resistant material that prevents water seepage into a building and directs the flow of moisture in walls

Module 45103   Review Questions

1. What is the main purpose of insulating a structure?
   a. To control the movement of heat through a wall
   b. To reinforce the structural members with additional stability
   c. To make a structure as soundproof as possible
   d. To finish the interior walls of the structure

2. The two overall categories of insulation are _____.
   a. continuous and rigid
   b. rigid and foam
   c. flexible and loose-fill
   d. cavity and continuous

3. Heat will always flow (or conduct) through any material or gas from _____.
   a. the interior to the exterior
   b. the exterior to the interior
   c. a higher temperature area to a lower temperature area
   d. a lower temperature area to a higher temperature area

4. Which type of insulation is specifically required by the *IECC*® to ensure sufficient insulation?
   a. Fiberglass insulation
   b. Reflective insulation
   c. Continuous insulation
   d. Lightweight aggregates

5. What is one place where insulation should be installed in a building?
   a. Garages
   b. Roofs
   c. In floors above the first floor
   d. In interior walls

6. You are installing insulation for a building in Juneau County, Alaska. In the *IECC*® it is marked as 7 Juneau. What does that indicate?
   a. It is in Climate Zone 7. Moisture regime is irrelevant.
   b. It is in Climate Zone 7 and is dry.
   c. It is in Climate Zone 7 and is moist.
   d. It is in Climate Zone 7 and is marine.

7. Wood or metal studs are considered thermal bridges because they _____.
   a. create considerable breaks in the insulation
   b. channel cold air out of the building
   c. prevent expansion and contraction of building materials
   d. do not support continuous insulation

8. Which materials are used to make vegetable (natural) insulation?
   a. Corn stalks and cotton
   b. Glass and cotton
   c. Straw and perlite
   d. Foil and wood

9. Which material can be made into foam insulation?
   a. Slag
   b. Cork
   c. Polyurethane
   d. Copper

10. Which of the following is the best insulator?
    a. 4" thick concrete block wall
    b. Plastic film
    c. 3.5" thick fiberglass batt
    d. Building paper

11. What can the vapor retarder on kraft-faced insulation be made of?
    a. Slag, phenolic, or copper
    b. Fiberglass or plastic
    c. Cork, tin plate, or paper
    d. Paper, vinyl, or foil

12. How should kraft-faced insulation be placed to prevent moisture from building up in the insulation?
    a. The vapor retarder face should be on the inside of the insulation stack.
    b. The vapor retarder face should be on the outside of the insulation stack.
    c. It should be on the center of a stack with a waterproof membrane on either side.
    d. It should be between the sheathing and exterior air barrier with the vapor retarder facing out.

13. Because reflective insulation reflects light, it can be used to protect against _____.
    a. heat in areas that get a lot of sunlight
    b. cold in areas with severe temperature drops
    c. moisture in areas with a lot of rain
    d. cold in areas with little sunlight

14. Sprayed-in-place insulation must be applied _____.
    a. only when the temperature is above 70°F
    b. by trained and certified contractors
    c. on top of loose-fill insulation
    d. by the contractor who installs the framing studs

15. Which type of insulation can be made of sintered clay products or cinders?
    a. Foamed-in-place insulation
    b. Lightweight aggregates
    c. Rigid expanded polystyrene
    d. Mineral wool insulation

16. Mineral wool board is made of _____.
    a. plastic
    b. polystyrene
    c. phenol-based synthetic polymer
    d. bonded rock fibers

17. What should be checked before installation to determine required R-values and the types of insulation required or permitted?
    a. Building plans and applicable codes
    b. The architect's plans
    c. Building blueprints
    d. The job summary analysis

18. Flexible insulation is used in _____.
    a. unfinished floors
    b. windows
    c. foundations
    d. porches

19. What type of insulation can you use to fill the narrow spaces around windows and doors?
    a. Rigid insulation
    b. Loose-fill insulation
    c. Foam insulation
    d. Reflective insulation

20. What is an acceptable height for the numbers labeling markers in loose-fill insulation?
    a. $\frac{1}{4}$"
    b. $\frac{1}{2}$"
    c. $\frac{3}{4}$"
    d. 1"

21. For loose-fill insulation, use a _____ to tamp the insulation and level it to the required depth for the R-value desired.
    a. straight edge
    b. marker
    c. shovel
    d. strike-through board

22. When installing loose-fill insulation, ensure that a separate _____ has been installed to prevent moisture and dust penetration.
    a. soffit baffle
    b. sheathing
    c. vent
    d. vapor retarder

23. Insulation below grade is required in _____.
    a. Climate Zones 1–3
    b. Climate Zones 4–8
    c. Climate Zones 2–6
    d. Climate Zones 3–5

24. Foam insulation can be applied to the exterior sheathing of a structure to create _____.
    a. continuous insulation
    b. thermal resistance
    c. water vapor
    d. diffusion

25. What does condensed moisture do to insulation?
    a. It enhances the protection of insulation.
    b. It reduces the efficiency of insulation.
    c. It causes insulation to freeze.
    d. It helps insulation reduce the building's heat loss.

26. Continuous insulation helps with moisture control by _____.
    a. blocking cold air from the building
    b. allowing hot air into the building
    c. sealing the gaps in the wall assembly
    d. allowing for movement and expansion of building materials

27. When a vapor retarder is used under the ceiling, proper free air ventilation for a gable roof is defined as 1 ft^2 for every _____ ft^2 of attic area.
    a. 150
    b. 160
    c. 300
    d. 320

28. When vapor reaches its _____, it condenses.
    a. permeability
    b. dew point
    c. convection
    d. U-factor

29. Class III vapor retarder is permitted for vented cladding over wood structural panels in which of the following climate zones?
    a. Zones 1 and 2
    b. Zones 3 and 4
    c. Marine 4 and Zone 5
    d. Zones 5 and 7

30. When installing a vapor retarder in a crawl space, each joint should be taped or sealed, and overlap by _____.
    a. 4"
    b. 6"
    c. 7"
    d. 8"

31. Fluid-applied water-resistive barriers should be applied only when the surface and air temperature is between _____.
    a. 60°F and 80°F
    b. 30°F and 70°F
    c. 70°F and 100°F
    d. 40°F and 100°F

32. The vertical seams of building wrap are usually overlapped _____ at each corner of the building.
    a. 2" to 6"
    b. 3" to 4"
    c. 6" to 12"
    d. 12" to 18"

33. What does an expansion joint do?
    a. It connects exterior building material to interior building material.
    b. It allows building materials to expand and contract without distress.
    c. It prevents the expansion or movement of building materials.
    d. It controls cracking in masonry caused by material shrinkage.

34. In general, water-resistive barriers should overlap by at least how many inches where joints occur?
    a. 2"
    b. 4"
    c. 6"
    d. 8"

35. Which of the following statements is true about the material used to seal a joint?
    a. It should be flexible.
    b. It should be rigid.
    c. It should be air permeable.
    d. It should be reflective.

Answers to Odd-Numbered Module Review Questions are found in *Appendix A*.

CORNERSTONE OF CRAFTSMANSHIP

Kevin Howser

Vocational Instructor
Federal Correctional Institution Cumberland

How did you choose a career in the industry?
Ever since I was a kid, I have enjoyed creating. As a teenager in school, I took a wood shop class that inspired me to set up a small workshop in a shed and make my own gifts and presents. That led to advanced woods, drafting, and then completion of a vocational building trades program in high school.

Who inspired you to enter the industry?
My Uncle Don was a big influence in my life. He was a master carpenter that worked out of the local carpenters union.

What types of training have you been through?
I took a vocational building trades class during high school and was an honor graduate. Our hands-on training consisted of building a house that was later sold to help support the program. I also received a Bachelor of Science degree in accounting and am a licensed Certified Public Accountant (CPA).

How important is education and training in construction?
Materials and methods in construction are always evolving. If you stop learning, the industry will pass you by. My degree in accounting and business helped me manage a contracting business.

How important are NCCER credentials to your career?
I wear a lot of hats in my current position (Sponsor Representative, Master Trainer, and Craft Instructor for CORE, Carpentry Level 1 and 2, and Drywall Level 1).

How has training/construction impacted your life?
I have used my training to make a living and personally. I have worked as a contractor, renovated investment properties, built and renovated my personal property, and recently built a house for my sister.

What kinds of work have you done in your career?
I have split my career between construction and accounting, auditing, tax, and finance work. I combined the two as a residential contractor and business owner for over 10 years. I have been teaching residential carpentry and drywall for over 7 years.

Tell us about your current job.
I teach CORE, Drywall L1, Carpentry Level 1 and 2, OSHA 10 for Construction and Forklift classes at a medium security federal prison.

What do you enjoy most about your job?
Providing the opportunity for students to change their lives. The construction field allows anyone with work ethic and drive the opportunity to earn a living wage. To use accounting terminology "to change a person from a liability to an asset."

What factors have contributed most to your success?
I am not afraid to work and learn new things.

Would you suggest construction as a career to others? Why?
Yes. I like the satisfaction of creating and completing projects. My kids still get on me for saying, "I built that house, I built that garage, I built that deck, and etc...."

What advice would you give to those new to the field?
I would give them the advice my grandfather gave me, "It doesn't matter what you do in life, but be the best you can be."

Interesting career-related fact or accomplishment:
Working with Habitat for Humanity on a house build by having students build interior and exterior wall sections to be shipped to the jobsite. Students took pride in the work they accomplished and were able to see their skills used in the community.

How do you define craftsmanship?
I would define craftsmanship as paying attention to detail and taking pride in what you do.

Answers to Section Review Questions

Answer	Section	Objective
Section One		
1. b	1.1.1	1a
2. d	1.2.1	1b
Section Two		
1. a	2.1.1	2a
2. b	2.2.0	2b
3. b	2.3.0	2c
4. d	2.4.2	2d
Section Three		
1. b	3.1.1	3a
2. a	3.2.1	3b
3. a	3.3.0	3c
4. c	3.4.1	3d
Section Four		
1. c	4.1.1	4a
2. d	4.2.0	4b
Section Five		
1. d	5.1.0	5a
2. a	5.2.2	5b
3. d	5.3.1	5c

Drywall Installation

Source: Clerik82/Shutterstock

Objectives

Successful completion of this module prepares you to do the following:

1. Identify and describe the materials and tools required to install drywall.
   a. List characteristics of gypsum board and explain why it is used.
   b. Describe types of drywall fasteners and explain how they are used.
   c. Identify and describe types of adhesives and explain how they are used for drywall work.
   d. Identify and describe tools used to install drywall.
2. Explain the procedures for installing gypsum board on different types of ceilings and walls.
   a. Describe the procedure for single- and multi-ply installations of gypsum board on ceilings and walls.
   b. Identify and describe the types of drywall trims.
   c. Explain how moisture is controlled to prevent its effects on walls and ceilings.
   d. Explain guidelines for fire-resistance-rated and sound-rated walls.

Performance Tasks

Under supervision, you should be able to do the following:

1. Install gypsum drywall panels on wood joist ceilings and wood stud walls using the following:
   - Nails
   - Screws
2. Install gypsum drywall panels on steel grid ceilings and steel stud walls using screws.

Overview

Gypsum board is the most common drywall finish used inside residential and commercial construction. A variety of gypsum materials are used for these applications, and a number of construction methods are used to build standard ceilings and walls, as well as assemblies that meet building codes for fire resistance and sound control. Gypsum board drywall installation requires the knowledge and use of specialized tools, fasteners, and construction methods dictated by building codes, and therefore must be carefully considered.

Digital Resources for Drywall

NCCER

SCAN ME

Scan this code using the camera on your phone or mobile device to view the digital resources related to this craft.

1.0.0 Materials and Tools

Performance Tasks

There are no Performance Tasks in this section.

Objective

Identify and describe the materials and tools required to install drywall.

a. List characteristics of gypsum board and explain why it is used.
b. Describe types of drywall fasteners and explain how they are used.
c. Identify and describe types of adhesives and explain how they are used for drywall work.
d. Identify and describe tools used to install drywall.

Watching experienced workers install gypsum board walls inside a building, you might think the job is easy. They work in pairs, lifting a long panel and holding it firmly against the studs while one of them places the nose of the screw gun against the board and pulls the trigger. They move quickly and easily, covering the studs with perfectly straight, flat rows of panels.

In reality, expert installation of gypsum board takes a good deal of knowledge and practice. Gypsum boards are heavy, and it is not as easy as it looks to install them straight and tight against the studs and ceiling joists and to cut out around electrical outlets. Additionally, each construction project requires you to take the time to learn and understand the specifications for that project and how to meet them, as well as the building codes for the area in which the construction is taking place.

You are prepared to install gypsum board when you understand the following:

- The characteristics of gypsum board
- The fasteners, tools, and materials used to install gypsum board
- How to work safely with gypsum board
- The methods for applying gypsum board to different surfaces
- How to build fire-resistance-rated and sound-rated assemblies with gypsum board

1.1.0 Gypsum Board

Gypsum board: A generic term for paper-covered panels with a gypsum core; also known as *gypsum drywall*.

Gypsum board, also known as *gypsum drywall* when used inside buildings, is one of the most popular and economical methods of finishing the interior walls and ceilings of wood-framed and steel-framed buildings. Gypsum drywall gives a wall or ceiling made from many panels the appearance of being made from one continuous sheet.

Gypsum board is a generic name for products made with the mineral gypsum that feature a noncombustible core and a paper covering on the face, back, and long edges of each board. A typical residential (horizontal) board application is shown in *Figure 1* and a typical commercial (vertical) board application is shown in *Figure 2*.

Figure 1 Typical residential (horizontal) board application.
Source: Rachid Jalayanadeja/Shutterstock

Figure 2 Typical commercial (vertical) board application.
Source: ungvar/Shutterstock

The main difference between gypsum board and other sheet products such as plywood, hardboard, and fiberboard is its natural ability to resist fire because of its gypsum content. Gypsum is a mineral, calcium sulphate dihydrate, found in sedimentary rock formations. Human-made synthetic gypsum is also used in gypsum board production, mixed with gypsum rock mined from the earth. Gypsum is naturally noncombustible because it contains water in its crystal molecular structure.

Gypsum drywall is generally available in widths of 4' and lengths of 8', 10', 12', and 14'. Other lengths are available by special order. The 4' side of a gypsum board, the side that is not covered in paper, is called the **end** of the board. The longer side, which is covered with paper, is called the **edge** of the board. The ends of gypsum panels are square, but panels are manufactured for different purposes with a variety of different edges, including rounded, tapered, beveled, and square (*Figure 3*).

End: The side of a gypsum board perpendicular to the paper-bound edge. The gypsum core is always exposed.

Edge: The paper-bound edge of a gypsum board as manufactured.

Calcination: The process of heating gypsum rock enough to evaporate most of the water in its molecular structure, causing a chemical change in the material.

The Way It Was

Until the 1930s, walls were typically finished by installing thin, narrow strips of wood or metal known as *lath* between studs and then coating the lath with wet gypsum plaster. Skilled plasterers could produce a smooth wall finish, but the process was time-consuming and messy. Paper-bound gypsum board was introduced in the early 1900s. After World War II, it came into widespread use as a replacement for the tedious lath-and-plaster process.

1.1.1 Advantages of Gypsum Board

Gypsum board walls and ceilings have the following outstanding advantages:

- *Fire resistance* — Gypsum board is an excellent fire-resistive material because the gypsum in its noncombustible core contains molecules of water. Under high heat, this water is slowly released as steam, which retards the flow of heat through the assembly. Even after complete **calcination**, when all the water has been released from the gypsum in the board, gypsum drywall continues to provide a barrier to the spread of flames. Tests conducted in accordance with the *ASTM Test Standard E84* show that gypsum drywall resists flame spread and creates little smoke during a fire.

- *Sound attenuation* — A key consideration in the design of a building is how to control unwanted sound transmission between and within rooms. Standard gypsum board applied to walls and ceilings, because of its inherent mass and elasticity, effectively helps to control sound transmission between and within rooms to a certain extent. But wall and ceiling assemblies that use standard

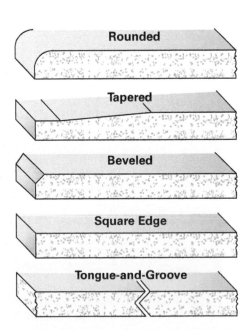

Figure 3 Types of edges of gypsum board.

gypsum board are non-rated for sound control. Enhanced gypsum board and specialized insulation are available for use in sound-rated assemblies.

- *Durability* — Walls and ceilings made from gypsum board are strong and have excellent dimensional stability. The wide range of gypsum products includes abuse-resistant and impact-resistant panels for high-traffic or high-contact settings. There are also gypsum board options that withstand damp conditions and resist moisture and mold.
- *Economy* — Gypsum board products are inexpensive to manufacture and relatively easy to apply, and so they may be installed at relatively low cost. They are the least expensive choice among wall surfacing materials that offer a fire-resistant interior finish.
- *Versatility* — Gypsum board products satisfy a wide range of architectural requirements. They are readily available and easy to apply, repair, decorate, and refinish. This combination of traits is unmatched by any other wall surfacing product.
- *Sustainability* — Gypsum board is considered a sustainable material because, in addition to naturally mined gypsum, it uses flue gas desulfurization (FGD) gypsum, a synthetic byproduct of coal-burning power plants. Additionally, gypsum board manufacturers have for decades used recycled **face paper** on their gypsum board products.

Face paper: The paper bonded to the surface of gypsum board during the manufacturing process.

Going Green

Recycling Gypsum Board Waste

The gypsum board waste that results from construction projects can be used for many purposes, including as a filler in plastics and cement, and as a soil conditioner that has been shown to reduce the levels of toxic soluble reactive phosphorus in waterways plagued by fertilizer runoff. With the proper separation of waste materials on the worksite—a practice that is growing in popularity—gypsum board without any fasteners or tape can be reclaimed by manufacturers to make new gypsum board. Leading gypsum board brands are now offering panels containing some recycled material. Companies such as USA Gypsum partner with building contractors to recycle gypsum board waste. Since 1998, USA Gypsum has recycled millions of pounds of gypsum board.

Gypsum Drywall—A Versatile Finish Material

It is common to think of gypsum drywall as a finish for flat walls and ceilings. These photographs show how it can be used in much more complex designs. Manufacturers have developed highly flexible gypsum boards, called Hi-Flex, to facilitate applications requiring complex designs.

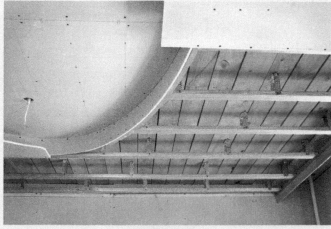

In Process

Source: Sever180/Shutterstock

Finished

Source: Jirawatfoto/Shutterstock

1.2.0 Drywall Fasteners

Nails and screws are commonly used to attach gypsum board to framing members in both **single-ply** (one-layer) **construction** and **multi-ply** (multiple-layer) **construction**. Special drywall adhesives can be used to secure single-ply gypsum board to framing, furring, masonry, and concrete, or to laminate a face ply to a base layer of gypsum board or other base material. Adhesives must be supplemented with mechanical fasteners, such as nails or screws.

Installing gypsum board requires that you understand how to **countersink** drywall nails and screws. Countersinking means driving fasteners into the board until the head of the fastener rests just below the surface of the face paper on the board, without breaking through it. The face paper on a gypsum board, bonded as it is to the gypsum core, is a vital part of the structure of the board, so take particular care to never break it when driving a nail or screw. A fastener that has broken through the face paper is not holding the board properly to the framing member. The gypsum core around the nail or screw is now likely to crumble, causing the board to loosen or even detach from the framing member. If you do break the face paper on a board while driving a nail or screw, drive another fastener properly near the same place.

When you countersink a nail or screw correctly, with the head seated just below the surface of the panel, you create a uniform depression (see *Figure 4*). This depression leaves room to apply **joint compound**, also called *mud*, during the **finishing** process, to mask the presence of the nail or screw and create a smooth surface.

All gypsum board fasteners must be used with the correct safety shield, tool guard, or attachment recommended by the manufacturer.

1.2.1 Nails

Both **annular nails** and **cupped-head nails** are acceptable for applying gypsum board to wood framing or wood furring strips (see *Figure 5*). Annular nails have rings around the shank, the long part of the nail, which is why they are also called *ring shank nails*. Annular nails grip wood studs and create good **withdrawal resistance**, compared to nails with smooth shanks. Cupped-head nails have concave heads designed to countersink properly into gypsum board. Any nails used to install gypsum board should have thin-rimmed flat or concave heads. The heads should be between $\frac{1}{4}$" and $\frac{5}{16}$" in diameter to provide adequate holding power without cutting the face paper when the countersunk nail creates the proper depression.

Single-ply construction: A wall or ceiling installation built with one layer of gypsum board.

Multi-ply construction: A wall or ceiling installation built with more than one layer of gypsum board.

Countersink: To drive a nail or screw through a gypsum board and into the framing member until the head of the fastener rests just below the surface, without breaking the face paper.

Joint compound: A mixture of gypsum, clay, and resin applied wet during the finishing process to the taped joints between gypsum boards to cover the fasteners, and to the corner bead and accessories to create the illusion of a smooth unbroken surface. Sometimes called *mud* or *taping compound*.

Finishing: The application of joint tape, joint compound, corner bead, and primer or sealer onto gypsum board in preparation for the final decoration of the surface.

Annular nails: Nails with rings around the shank, which provide a stronger grip and higher withdrawal resistance than nails with smooth shanks.

Cupped-head nails: Nails with a concave head (shaped like a cup) and a thin rim.

Withdrawal resistance: The amount of resistance of a nail or screw to being pulled out of a material into which it has been driven.

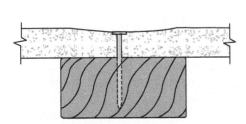

Figure 4 Uniform depression.

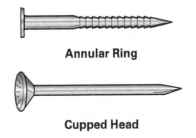

Annular Ring

Cupped Head

Figure 5 Nails.

Nails must be long enough to go through the gypsum board and far enough into the supporting wood construction to provide adequate holding power. The nail penetration into the framing member should be at least $\frac{7}{8}$" for smooth shank nails and $\frac{3}{4}$" for annular nails, which provide more withdrawal resistance and therefore require less penetration. For fire-resistance-rated assemblies, greater penetration is required (generally $1\frac{1}{8}$" to $1\frac{1}{4}$" for one-hour assemblies). Follow the specifications in the fire test report when choosing nails for a fire-resistance-rated assembly.

Drywall nails should be driven into gypsum board with a hammer specifically designed for installing gypsum board, such as the hammer end of a

CAUTION

Casing and common nails have heads that are too small in relation to the shank; they easily cut into the face paper and threaten the integrity of the board and its attachment to the stud. Nail heads that are too large are also likely to cut the paper surface if the nail is driven incorrectly at a slight angle.

CAUTION

Never try to drive drywall nails into gypsum board with a standard hammer, which is not shaped properly to countersink drywall nails and can much more easily break the face paper.

Tooth: A textured surface created mechanically to help a covering material adhere more effectively.

Thread: The protruding rib of a screw that winds in a helix down its shank.

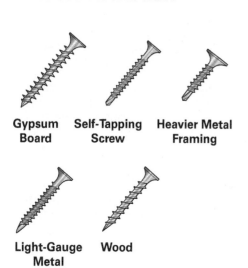

Gypsum Self-Tapping Heavier Metal
Board   Screw   Framing

Light-Gauge Wood
Metal

Figure 6 Drywall screws.

Furring strip: A flat, narrow piece of wood attached to wood framing to create a level surface in preparation for installing gypsum board.

Pitch: The number of threads per inch on a screw shank, with more threads producing a finer pitch and fewer threads producing a coarser pitch.

Self-tapping screw: A screw that bores an internal screw thread in the material into which it is driven, creating a strong hold between the material and the screw.

Furring channel: A long, narrow piece of metal bent into the shape of a hat (which is why it is also called a *hat channel*), with two flanges (the brim of the hat) on either side of a channel (the crown of the hat), used to create a level surface on uneven masonry or metal framing in preparation for installing gypsum board.

Self-piercing screw: A screw with a point sharp enough to penetrate steel.

drywall hatchet, or with another hammer with a special wide curved head that forms the required depression around the nail head without breaking the face paper. The head of a drywall hammer has a waffle texture, which transfers to the depression around the nail head. This textured surface, or **tooth**, helps joint compound adhere to the depression during the finishing phase. Particular care should be taken when driving nails into gypsum board not to crush the core by striking it too hard with the hammer. Nails are not an appropriate fastener for applying gypsum board to steel framing.

Special Fasteners

Application of gypsum board requires special fasteners. Common nails and ordinary wood or sheet metal screws are not designed to penetrate the board without damaging it, or to hold the board tightly against the framing, or to permit correct countersinking for proper concealment of the fasteners during finishing.

1.2.2 Screws

Drywall screws (*Figure 6*) are used to attach gypsum board to wood or steel framing or to other gypsum board. They have a pronounced **thread** and a Phillips head design that is intended to be used with a drywall screw gun. The bugle-shaped head of a drywall screw is specially designed to countersink into the gypsum board and pull it tightly against the framing member. When a drywall screw is properly driven into a gypsum board, it creates a uniform depression that is free of ragged edges and fuzz.

When choosing the right screws for attaching gypsum panels to framing members, your main consideration will be what type of material you are attaching them to, as shown in *Table 1*.

TABLE 1 Drywall Screw Types

Threads	Application
Coarse	Drywall to wood
Fine, self-tapping	Drywall to steel: 18 mil (25 gauge) or 27 mil (22 gauge)
Fine, self-drilling	Drywall to steel: 30 mil (20 gauge)
Coarse	Drywall to drywall

The screw threads determine which of the following applications they are best suited for:

- A coarse screw is designed to fasten gypsum board to wood framing or a wood **furring strip**. This type of screw has a coarse **pitch**, which means the threads are relatively far apart. The point is diamond-shaped to pierce both gypsum board and wood, and the coarse threads hold the gypsum panels securely to the wood framing members.

- A fine, **self-tapping screw** is designed to fasten gypsum board to a standard 18 mil (25 gauge) or 27 mil (22 gauge) steel stud or steel **furring channel**. It has fine, close threads, so there are more of them on the shank of the screw. This type of drywall screw is a **self-piercing screw**, which means it is made with a sharp tip designed to easily pierce steel framing members. Easy penetration is important, because steel studs are often flexible and tend to bend away from the screws, and the screws tend to strip easily. These screws are also self-tapping: as they bore into the steel framing member, they tap the steel stud, which means they carve internal screw threads into the steel, creating a tight hold between the screw and the steel framing member similar to that between a nut and a bolt.

- A fine, **self-drilling screw** is recommended when attaching gypsum panels to steel framing that is 30 mil (20 gauge). The tip of this type of screw is shaped like a drill bit for cutting through thicker steel studs with ease. Self-drilling screws are also self-tapping.

- A coarse screw with a deeper, special thread design to create a tight grip between boards is used for fastening gypsum board panels to gypsum backing boards in multi-ply construction. The most common length of screws used for multi-ply construction is $1\frac{1}{2}$", but other lengths are available. However, in two-ply construction, in which the face layer is screw-attached to the base layer, additional holding power is developed in the base layer, which permits a reduced penetration in the face layer of a minimum of $\frac{1}{2}$".

Self-drilling screw: A screw with a point shaped like a drill bit to penetrate steel.

Did You Know?

Steel Framing Member Size

The thickness of the steel framing member determines its use. For drywall applications, the size of the steel framing members also determines the type of fasteners needed.

Thickness, in Millimeters	Reference Gauge No.
18	25
27	22
30	20 Drywall
33	20 Structural
43	18
54	16
68	14
97	12
118	10

Source: Adapted from the Steel Framing Industry Association (SFIA) specifications: https://sfia.memberclicks.net/assets/TechFiles/SFIA%20Tech%20Spec%202015%20updated%207.12.17%20v.2.pdf

The thickness of the gypsum board you use will depend on the building codes and the project specifications. The most common board thicknesses for single-ply interior wall assemblies are $\frac{1}{2}$" and $\frac{5}{8}$". For ceilings, many installers use $\frac{5}{8}$" board exclusively, because other thicknesses tend to sag or ripple over time. The correct screw length is based on the thickness of the gypsum panels and the framing material. For example, the recommended minimum penetration into wood studs for single-ply construction is $\frac{5}{8}$". That means if you are attaching $\frac{5}{8}$" board, you need screws that are at least $1\frac{1}{4}$" long.

Keep in mind that the required fastener penetration changes when constructing fire-resistance-rated assemblies with steel framing, in which case nails or screws must be at least 1" long and may need to be specially treated. Always read the architectural specifications and local building codes for an individual project and follow them carefully.

Selecting Drywall Screws

It is important to use the correct screws for each job. The numbers on the box will indicate the diameter (6, 8, 10, etc.), the length ($1\frac{1}{4}$", $1\frac{5}{8}$", etc.), and the thread (coarse or fine). A #6 screw, which has a relatively small diameter, is appropriate for applying gypsum board to wood and steel framing in single-ply construction and is the most common size screw used by gypsum board installers in the field.

Specialty panels may require specialty fasteners. Follow manufacturers' guidelines for installing specialty products.

1.3.0 Adhesives

Adhesives are used to bond single layers of gypsum board directly to the framing, furring, masonry, or concrete. They can be used to laminate gypsum board to base layers of backer boards, sound deadening boards, rigid foam, and other rigid insulating boards. The adhesive must be used in combination with screws, which provide supplemental support.

When choosing an adhesive, either to apply gypsum board directly to framing members or to apply a face layer to a base layer in a multi-ply assembly,

NOTE

Check the temperature range for the adhesive you plan to use to make sure it is compatible with the expected operating temperatures of the building.

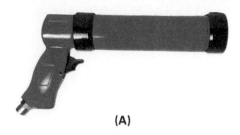

(A)

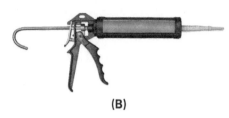

(B)

Figure 7 Adhesive applicators.
Sources: (A) darksoul72/Shutterstock;
(B) donatas/1205/Shutterstock

Bead: An application of adhesive or other construction material in a sphere or line not less than $\frac{3}{8}$" in diameter.

make sure it is recommended by the manufacturer for the specific conditions of your project. There are many all-purpose drywall adhesives available, but some of them may not be effective to hold gypsum board tightly to all types of framing members.

Whichever drywall adhesive you choose, follow all manufacturer directions for use. If the adhesive has a solvent base, follow appropriate precautions to keep the jobsite safe. Do not use solvent-based adhesives near an open flame or in poorly ventilated areas.

The most common type of drywall adhesive is applied with an electric, pneumatic, or hand-operated caulking gun (*Figure 7*) in a continuous or semicontinuous **bead**.

Drywall contact adhesive is a type of adhesive that can be used to apply a face layer of gypsum board to a base layer in multi-ply installations. It can also be used to apply gypsum board to steel studs. Contact adhesive is applied by roller, spray gun, or brush in a thin, uniform coating to both surfaces to be bonded. For most contact adhesives, some drying time is usually required before surfaces can be joined and the bond can be developed.

WARNING!

Observe all safety data sheet (SDS) precautions for adhesives. Extreme caution must be taken when using contact cement because it is highly flammable. It also produces toxic fumes and should be used in a well-ventilated area, as the fumes can quickly overcome a worker. Always wear appropriate PPE and follow manufacturer's instruction when using adhesives.

Figure 8 T-square.
Source: Photo Win1/Shutterstock

Figure 9 Utility knife.
Source: Andrei Kuzmik/Shutterstock

1.4.0 Tools Used to Install Gypsum Board

The following tools are commonly used to install gypsum board:

- 4' T-square — This tool is indispensable for making accurate cuts across the narrow dimension of gypsum board. See *Figure 8*.
- *Utility knife* — This is the standard knife used for cutting gypsum board. It has replaceable blades stored in the handle. Knife blades need to stay sharp to cut the gypsum boards without needing to apply too much pressure. See *Figure 9*.
- *Rasp* — The rasp (*Figure 10*) is used to quickly and efficiently smooth rough-cut edges of gypsum board. The tool has both a file for finishing and a rasp for rough shaping.
- *Circle cutter* — The circle cutter (*Figure 11*) has a calibrated steel shaft that allows accurate cuts up to 16" in diameter. The cutter wheel and center pin are heat-treated.
- *Drywall saw* — This saw, also known as a *keyhole saw*, is used for cutting small openings and making odd-shaped cuts (*Figure 12*). A power cutout

Figure 10 Drywall rasp.
Source: malaha/Shutterstock

Figure 11 Circle cutter.
Source: Dmitry Naumov/Shutterstock

Figure 12 Drywall saw.
Source: Stocksnapper/Shutterstock

tool, also called a *drywall router*, (*Figure 13*) can be used for the same purpose. The power cutout tool comes with a depth gauge you can set before cutting, thereby ensuring that you cut only through the gypsum board. The depth gauge can help beginners get used to how the power cutout tool works, but more advanced drywall installers often prefer to work without it.

Figure 13 Power cutout tool.
Source: Keith Homan/Shutterstock

Measuring and Marking

When measuring gypsum board, use a soft lead pencil to mark the panels. Marks made by a ballpoint pen or marker may bleed through the joint compound and paint.

- *Gypsum board foot-operated lifter* — A foot-operated gypsum board lifter slides under a gypsum panel being installed just above floor level. When you press down on the lever with your foot, it lifts the panel up and pushes it forward into place so that you can easily attach it to the framing members. The lifter can be used for either parallel or perpendicular board applications.

- *Drywall hatchet* — This tool includes a drywall hammer with a wide symmetrical convex head designed to compress the gypsum panel face and leave the desired depression without breaking the face paper. The blade end can be used to cut boards roughly down to size, but then the ends must be smoothed before the boards are applied to the wall assembly. See *Figure 14*.

- *Screw gun* — Electric drywall screw guns, such as those shown in *Figure 15*, are designed to drive steel screws through gypsum board and into the framing member to a precise depth. The best screw guns for installing drywall on wood or steel studs have powerful brushless (meaning almost frictionless) motors that run at a minimum of 4,000 revolutions per minute (rpm). Screw guns feature an adjustable depth setting, located on the nose cone of the tool, for use with gypsum board of different thicknesses. When the screw being driven into the board reaches the preset depth, a clutch mechanism in the screw gun disengages and stops the screw from being driven any deeper. Manually loading a single-shot screw gun is as easy as placing the head of a drywall screw onto the magnetic bit tip, which will hold the screw in place until you press the screw against the board in the preferred location and pull the trigger. Many screw gun models offer a **collated magazine** that attaches to the nose cone and automatically feeds collated screws—screws arranged side by side on a long plastic strip—through the gun, allowing the installation process to move more quickly and easily.

- *Drywall lift* — This special device is designed to raise and support drywall panels during ceiling or high wall installations.

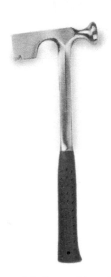

Figure 14 Drywall hatchet.
Source: Dorling Kindersley ltd/Alamy Stock Photo

Collated magazine: An attachment for a screw or nail gun that automatically feeds collated fasteners (fasteners arranged side by side on a strip of plastic) into the chamber of the gun for quick application.

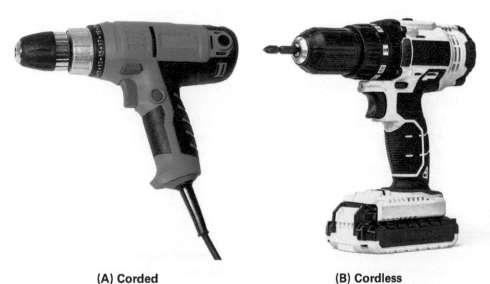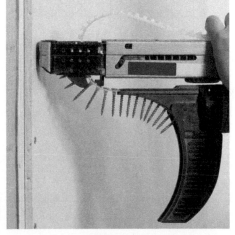

    (A) Corded   **(B) Cordless**   **(C) Collated**

Figure 15 Screw guns.
Sources: (A) krolya25/Shutterstock; (B) Nakornthai/Shutterstock; (C) Ilja Enger-Tsizikov/Alamy Stock Photo

NOTE

Attaching gypsum boards properly to framing members takes attention to detail. For best results, hold the screw gun perpendicular to the work surface. Adequate pressure must be exerted to engage the clutch and prevent the screw from slipping (also known as walking). The gun should be triggered continuously until a fastener is seated or sunk through the gypsum board and sufficiently into the framing member. A one-piece socket makes driving easier and more efficient than a separate socket and extension pieces, because it provides a more rigid base and firmer control.

1.0.0 Section Review

1. Because of its contents, _____ has a natural ability to resist fire.
   a. plywood
   b. hardboard
   c. gypsum board
   d. fiberboard

2. Fine, self-tapping screws are used to apply _____.
   a. drywall to standard 18 mil (25 gauge) or 27 mil (22 gauge) steel
   b. drywall to wood
   c. drywall to 30 mil (20 gauge) steel framing
   d. drywall to drywall

3. The most common type of drywall adhesive is applied with a _____ in a continuous or semicontinuous bead.
   a. caulking gun
   b. sprayer
   c. brush
   d. roller

4. A _____ comes with a depth gauge you can set before cutting to ensure you cut only through the gypsum board.
   a. circle cutter
   b. drywall hatchet
   c. power cutout tool
   d. utility saw

2.0.0 Installation

Performance Tasks

1. Install gypsum drywall panels on wood stud walls and ceilings using the following:
   • Nails
   • Screws
2. Install gypsum drywall panels on steel grid ceilings and steel stud walls using screws.

Objective

Explain the procedures for installing gypsum board and finishing different types of ceilings and walls.

a. Describe the procedure for single- and multi-ply installations of gypsum board on ceilings and walls.
b. Identify and describe the types of drywall trims.
c. Explain how moisture is controlled to prevent its effects on walls and ceilings.
d. Explain guidelines for fire-resistance-rated and sound-rated walls.

Any successful gypsum board installation project begins with a set of plans that dictate the materials and methods you will use to complete the project. Before you begin the gypsum board installation phase of the project, make sure you understand the project specifications and standards that apply to your portion of the project.

The responsibility for installing and finishing gypsum drywall varies from job to job and from one locale to another. In some situations, carpenters install, tape, and mud the drywall, and painters apply primer and surface **decoration**, such as paint or texturing. In other situations, professional drywall workers complete the installation, finishing, and decoration. The smaller the project, the more likely it is that the carpenter will install, finish, and decorate the drywall. Especially when you are first learning to install gypsum board, work with at least one other person to measure, cut, lift, and secure the panels tightly to the framing members.

Decoration: The application of the final surface covering on gypsum board walls.

2.1.0 Installing Gypsum Board

Gypsum board panels can be applied over any firm, flat base such as wood or steel framing or furring. Gypsum boards can also be applied to masonry and concrete surfaces, either directly or to surfaces that have been furred out with wood furring strips or steel furring channels.

The most common type of interior wall and ceiling assembly is the standard gypsum board system, which consists of a series of panels affixed with drywall nails or screws to wood or steel framing members. Installation of the boards leaves a series of vertical and horizontal **joints** between the gypsum panels. A butt joint is created where the short ends of two sheets of wallboard meet. A flat joint is the intersection of two bevel-edged wallboards. Once the panels have been attached to the framing members, the **outside corners** are reinforced with trim. Then the wall is finished, meaning the nail or screw depressions, the **inside corners**, and the joints between the panels are covered with **joint tape** and joint compound. Joint compound is applied to certain types of drywall trim as well.

Joints: Places where two pieces of wallboard meet.

Outside corners: Locations where two walls meet and face away from each other.

Inside corners: Locations where two walls meet and face each other.

Joint tape: Wide tape applied to the joints between gypsum boards and then covered with joint compound during the finishing process.

2.1.1 Single-Ply and Multi-Ply Construction

In structures that are not required to be fire-resistance-rated or sound-rated, single-ply gypsum board systems are commonly used (*Figure 16*). Multi-ply systems, as shown in *Figure 17*, have two or more layers of gypsum board to increase sound control and fire-resistive performance.

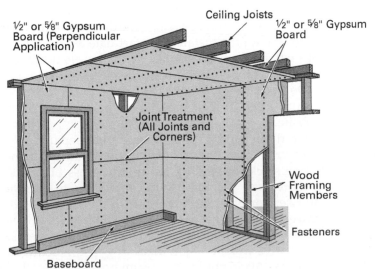

Figure 16 Single-ply construction.

The surface quality of multi-ply assemblies is often smoother than that of single-ply assemblies because face layers are often adhered to base layers, which means that fewer nails or screws are required to hold the face layer firmly in place. Additionally, the base layer of gypsum board reinforces the joints of the face layer, and the face layer masks any imperfectly aligned joints in the base layer. Because the face layer in a multi-ply assembly often requires fewer mechanical fasteners, problems with fastener pop occur less frequently.

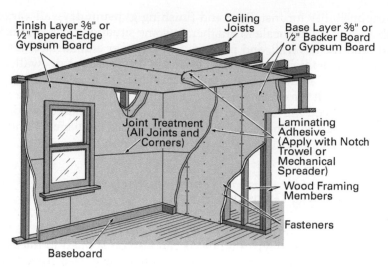

Figure 17 Multi-ply construction.

Satisfactory results can be assured with either single-ply or multi-ply assemblies by requiring the following:

- Proper framing details, consisting of straight, correctly spaced, and properly cured lumber
- Proper job conditions, including controlled temperatures and adequate ventilation during application
- Proper measuring, cutting, aligning, and fastening of the wallboard
- Proper joint and fastener treatment
- Special requirements for proper sound isolation, fire resistance, thermal properties, or moisture resistance

2.1.2 Jobsite Preparation

Materials for any gypsum board installation project should arrive at the jobsite in the manufacturer's original sealed packaging. The delivery of the board itself should coincide as closely with the installation start date as possible. Gypsum board must be protected from direct contact with the elements, including sunlight and precipitation. If the gypsum board is stored outside, it must be stored under cover. Do not store gypsum board directly on the ground. All materials should remain stored in their original wrappers or containers until ready to use on the jobsite.

Because gypsum board can bend if stored improperly, keep unopened packages stacked flat wherever you are storing them. There should be no more than 40 to 100 boards in one stack, depending on the thickness of the boards. If the project requires more than 100 boards, it is acceptable to pile the stacks in tiers separated by risers or other supports that are at least 4" wide. Make sure the supports between the tiers are carefully aligned from bottom to top so that each tier rests on a solid bearing rather than on the boards themselves, as shown in *Figure 18*. Excessive weight on any stack of gypsum board can cause individual boards to break. Avoid stacking gypsum boards vertically, because the heavy boards are not stable in this position and can cause serious injury if they fall.

Avoid stacking long lengths of gypsum board on top of short lengths to prevent the longer boards from bending or breaking. When moving boards, carry them or transport them with a dolly. Never drag gypsum boards along the ground or floor, which can damage the edges and make taping and mudding the joints more challenging. Avoid leaning boards against framing members for a prolonged period of time or during periods of high humidity, as the boards can warp.

Storing and Handling Gypsum Drywall

Figure 18 shows gypsum drywall as it would be stored in a warehouse or building supply store. Just before installation, the drywall panels would be distributed along interior walls and stood on their long side, or edge.

Drywall is heavy. Two $\frac{1}{2}$" × 4' × 8' panels weigh about 110 pounds, while a pair of $\frac{5}{8}$" × 4' × 8' panels weigh close to 150 pounds. The panels need to be handled carefully so they don't break under their own weight. They should be lifted and carried by the edges rather than the ends. Individual panels are heavy and awkward to lift, so it is important that you use proper lifting procedures when handling them in order to prevent serious back injury. That means keeping the board close to your body and lifting with your legs rather than your arms or back.

Figure 18 Gypsum board storage.
Source: Minute of love/Shutterstock

Construction materials respond to the environment around them. Job conditions such as temperature, humidity, and airflow can affect the performance of gypsum panels and joint treatment materials, the appearance of the joint, and the behavior of adhesive and finishing materials after project completion. While there are types of specialty gypsum board that are moisture and mold resistant, standard gypsum board is not appropriate for use in damp or wet conditions.

Ideally, the temperature at the worksite where you are installing gypsum boards will be around the same temperature at which the site will normally be kept once the work is completed. If it is not possible to keep the temperature at that level, then the site must be kept at a minimum of 40°F while you are installing the board mechanically, meaning with nails or screws. The boards themselves should be kept in the area where they will be installed for at least 48 hours before installation begins, so that they can acclimate to the environment. Gypsum boards not stored under these conditions may shrink or swell after installation causing cracks, ridges, and indentations.

If you are using adhesive to attach the gypsum board to the framing members or to a base layer of gypsum board, the temperature must be at least 50°F at the worksite for 48 hours before beginning the work and until the adhesive has dried completely. It must also be at least 50°F for 48 hours before finishing and decorating drywall and after completion until the joint compound, primer, or paint has dried completely. Whether you are applying gypsum board mechanically or with adhesive, the temperature at the worksite should never exceed 95°F.

Before beginning installation, review the project specifications and gather all the tools and materials you will need. Maintain a safe worksite by keeping pathways clear of materials and tools, and by cleaning up debris at the end of each workday. Wear proper personal protective equipment (PPE): Gypsum board is heavy, so you will need work boots to protect your feet from dropped boards. You will also need work gloves for lifting gypsum boards and to protect your hands against cuts when you are scoring boards with a utility knife. Installing gypsum board ceilings requires that you wear a hard hat to protect your head from falling boards and safety glasses to protect your eyes from dust and debris. If you are applying gypsum board with adhesive, maintain proper ventilation at the worksite and review SDS sheets for additional precautions.

2.1.3 Planning Your Installation

Any gypsum board installation should be carefully planned. The project specifications will be your guide to ensure a successful drywall installation.

Moving Drywall

A drywall dolly like the one shown here is specially designed to transport drywall panels on the jobsite. When loaded, always use two people to move a drywall dolly. Do not overload the dolly.

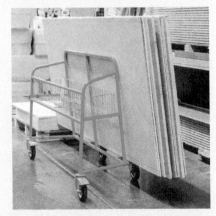

Source: Dmitry Markov152/Shutterstock

When installing gypsum board on both ceilings and walls, you will always install board on the ceilings first.

Before you begin installing the drywall, verify the following information with the project specifications:

- Check that all insulation and other internal features of the wall and ceiling assemblies have been installed, keeping in mind that some insulation is applied inside ceilings after gypsum board installation.
- Make sure that the spacing of the framing members is appropriate for the thickness of the gypsum board you are applying. Standard spacing for framing members in wall and ceiling assemblies is 16" **on center (OC)** or 24" on center. Generally, the thicker the gypsum board, the farther apart the framing members can be and still provide sufficient support. Effective support lessens the possibility that the panels will sag, crack, or ridge in the finished assembly.
- Determine the size of the gypsum board you will be using.
- Determine whether your application will be a **perpendicular installation** (horizontal) or a **parallel installation** (vertical).

ASTM C840, Section 9 designates the industry standard for single-ply drywall installations (*Table 2*) that you will see in the project specifications.

On center (OC): The distance between the center of one framing member or fastener to the center of an adjacent framing member or fastener.

Perpendicular installation: Applying gypsum board so that the edges are at right angles to the framing members, meaning the board is oriented horizontally.

Parallel installation: Applying gypsum board so that the edges are parallel to the framing members, meaning the board is oriented vertically.

TABLE 2 Single-Ply Maximum Framing Spacing

Gypsum Board Thickness	Application	Framing Members On-Center Spacing
Ceilings		
$3/8$"	Perpendicular	16"
$1/2$"	Parallel	16"
$5/8$"	Parallel	16"
$1/2$"	Perpendicular	24"
$5/8$"	Perpendicular	24"
Walls		
$3/8$"	Perpendicular or Parallel	16"
$1/2$"	Perpendicular	24"
$5/8$"	Parallel	24"

For multi-ply construction on ceilings and walls, there are many possible combinations of board thickness, application direction, and fastener spacing when considering the base ply and the face ply. These details differ based on framing member spacing and whether mechanical fasteners, adhesives, or both are being used between the face ply and the base ply. Check the project specifications to make sure you have the correct materials. If the project plans do not address specific details, consult the tables in *ASTM C840, Section 9* for guidance. These ASTM guidelines apply to multi-ply construction for both wood and steel stud framing.

When you apply gypsum board to ceilings, there is always the risk that the boards will sag after installation. To protect against this risk, make sure that the gypsum board is completely dry and has been allowed to reach the temperature of the worksite. Control the humidity at the worksite carefully, especially if you are using portable heaters, which tend to make the air more humid. One way to control humidity is to provide good ventilation on the worksite.

Once you have determined which boards you will use on the ceilings and walls, measure the framing space where you will be applying the gypsum board. Accurate measuring will usually reveal any irregularities in framing or furring. Poorly aligned framing should be corrected before applying the panels.

Choose the orientation of the boards—perpendicular or parallel—that meets project specifications, professional standards, and building codes and that

NOTE

Gypsum board measuring $1/4$" thick is not appropriate for single-ply application on wood or steel framing for ceilings or walls, but it can be used as a base ply or a face ply in multi-ply construction. It can also be applied directly to an existing surface of wood paneling, plaster, concrete, or masonry.

results in the fewest joints possible in the completed assembly. Joints between board ends and edges must be finished with tape and joint compound, so the fewer joints in an assembly, the less work it will take to finish the walls.

When ceilings are 8' high, perpendicular installation is generally preferred when using gypsum board with 4' ends, because two boards applied horizontally fit the height of the wall with little or no cutting to fit. In new construction, 9' ceilings have become the standard, and 54" wide gypsum panels are now more widely available, so that two boards applied horizontally fit the height of the wall with little or no cutting.

For rooms with ceilings taller than 9', parallel application is more practical and results in fewer joints. For certain projects, parallel application may be required for 8' ceilings in order to meet fire ratings.

For ceiling application, select the method that results in the fewest joints between boards. When ceilings are to receive water-based spray texture finishes, pay special attention to the spacing of framing members, the thickness of the board used, ventilation, vapor barriers, insulation, and other factors that can affect the performance of the system and cause problems, particularly sag of the gypsum board between framing members.

Parallel Versus Perpendicular Installation

Perpendicular installation often has the following advantages over parallel installation:

- There are fewer joints.
- Less measuring and cutting are required.
- Joints are at a convenient height for finishing.
- A single panel ties together more framing members to increase the strength of the assembly and hide framing irregularities.
- Panels applied perpendicularly are less likely to sag after finishing.

Plan to keep joints between boards at least 12" away from doors, windows, electrical outlets, and other openings in the assembly. Also plan to leave a gap of at least 1/4" at the bottom of the wall assembly, because standard gypsum board is not water or mold resistant and can absorb moisture held by concrete or other masonry. Placing a shim in the gap between the wallboard and the floor can also prevent the board from wicking moisture from the floor. Leave a gap anywhere else the gypsum board wall assembly abuts masonry materials.

Control Joints

In certain settings, it is necessary to install **control joints**, also called *expansion joints*, in wall and ceiling assemblies. Control joints are small gaps left between gypsum panels in an assembly to help prevent damage to the boards under everyday conditions, including the expansion and contraction of the boards that come with temperature fluctuations and the shifting of the boards that comes with normal building movement. Movement of the structure can impose severe stresses and cause cracks, either at the joint between boards or in the **field**, or inner area, of a board. U- or V-shaped inserts—often called *control joints, control strips,* or *expansion joints*—fill in the gaps between gypsum panels and allow for the desired smooth finished surface. See *Figure 19*.

When applying gypsum board to long stretches of framing, professional standards require that you create a control joint every 30' in wall partitions and 30' in ceiling assemblies, not to exceed 900 ft² between control joints without **perimeter relief**. In ceilings with perimeter relief, control joints are necessary every 50', not to exceed 2,500 ft² between control joints. In an assembly in which ceiling framing members change direction, either control joints or intermediate blocking is required. Without control joints, gypsum panels applied in an uninterrupted straight plane for long stretches are vulnerable to **centerline cracking**.

Control joints are appropriate in other locations as well. They should be installed over window and door openings, which are weak points in a structure

Control joints: Deliberate gaps left between long stretches of gypsum panels to allow them to expand, contract, or shift without cracking.

Field: The inner area of a gypsum panel.

Perimeter relief: A gap left between a ceiling assembly and a wall assembly to keep the two assemblies separate and to allow the gypsum board in the ceiling assembly to move freely.

Centerline cracking: A crack in a finished drywall joint that can occur as the result of environmental conditions or poor workmanship.

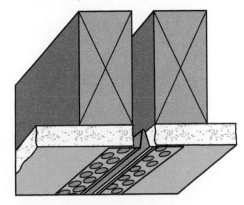

Figure 19 Typical control joint.

and areas of greater stress. Use control joints, metal trim, or other means to isolate gypsum panels from structural elements such as columns, beams, and load-bearing interior walls, and from dissimilar wall or ceiling finishes that might move differently. During planning, always confirm control joint locations with the architect or designer.

The modern trend toward less rigid structures in high-rise and commercial buildings has created a challenge for gypsum board installers. Because the structural members in these buildings naturally flex, expand, and contract more under everyday conditions, more of the load can be passed along to nonbearing walls and lead to drywall cracking and **ridging**. If you work on such a project, detailed designs are available for perimeter relief of nonbearing partitions to mitigate these risks. One solution is to use relief runners to attach nonbearing walls to ceiling and column members (*Figure 20*).

Ridging: A defect in finished gypsum board drywall caused by environmental or workmanship issues.

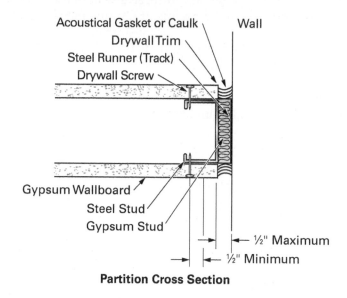

Partition Cross Section

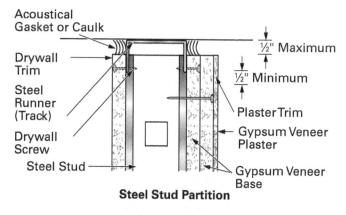

Steel Stud Partition

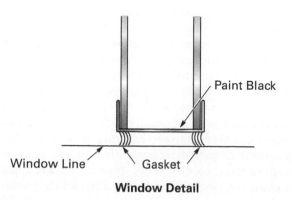

Window Detail

Figure 20 Designs for perimeter relief.

Building plans supplied by the project architect or designer should specify if and where control joints are to be incorporated in a wall or ceiling assembly. Installing control joints during the initial gypsum board installation will keep you—or someone else—from having to install them later, when cracks develop. If cracks do develop in a completed gypsum board installation in new construction, it is wise to wait until at least one heating season has passed before repairing or refinishing. During planning, always confirm control joint locations with the architect or designer.

Control Joints

Control (expansion) joints are used in large expanses of wall or ceiling drywall to compensate for the natural expansion and contraction of a building. Control joints help prevent cracking and joint separation. They are common in commercial construction, especially where exterior concrete walls contain expansion joints.

    If control joints are called for in a structure where fire-resistance rating and/or sound control are important, a seal must be installed first where the control joint will occur in the wall or ceiling. The control joint has a $1/4$" slot that is covered by plastic tape. The tape is removed after the joint is finished, leaving a small recess.

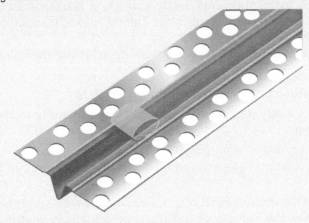

2.1.4 Cutting and Fitting Procedures

It is crucial to a gypsum board installation project that you measure, cut, and fit the boards accurately. Beginning with the ceiling panel installation, mark cutting lines on the panels using a 4' T-square and a pencil, such as a carpenter's pencil. Do not use a pen or a marker, which can bleed through primer applied during the finishing phase and cause project delays.

Step 1   Cut each board by first scoring through the paper down to the core with a sharp utility knife, working from the face side.

Step 2   Snap the board back, away from the cut face.

Step 3   Cut through the back paper with your utility knife. Less commonly, workers use a drywall saw to cut gypsum board.

Step 4   Cut away any jagged edges or burrs from the board with your utility knife. Use a rasp to smooth all cut edges and ends. Smooth ends and edges are vital to form neat, snug joints when the boards are installed. If burrs on the cut ends are not removed, they will form a visible ridge in the finished surface.

The wall or ceiling assembly you are building will include fixtures, such as electrical boxes, pipes, vents, and light fixtures, that must be marked and cut out of the gypsum board you are applying. As a beginning installer, take special care to mark and score the locations of these fixtures accurately on both the face and the back of the board before cutting out with a drywall saw or power router. Failure to do so can result in cutting mistakes that force you to scrap boards or to remove and replace boards you have already attached to the framing members.

The following are different ways to approach cutting holes in the panels for these fixtures:

- Some installers, especially more experienced workers, prefer to measure the location of a fixture with a measuring tape, mark the outline of the fixture on the board, cut it out with a utility saw or router, and then apply the board to the framing members.
- Other installers prefer to mark the location of the fixture on the board, attach the board to the framing members, and then cut out the space for the fixtures. The preferred method is to first tack the board to the framing members just enough to keep it in place, using a few fasteners at least 24" away from the cutout location. Do not screw the board tightly to the members until after you have completed the cutouts, because the pressure caused by the utility saw or router on a tightly attached board will break the board.

Marking cutouts accurately on gypsum panels can be challenging, and some experts recommend methods other than simply measuring with a measuring tape and marking boards with a pencil. This approach, if not done with care, can lead to mistakes and wasted boards.

Whichever marking and cutting method you use to accommodate wall and ceiling fixtures, make your cuts neatly and use a rasp to smooth the edges. Leaving rough edges and burrs on your cutouts makes it tricky to finish for a smooth surface.

To ensure a sound installation of gypsum drywall, use the following general practices:

- Install the ceiling boards first, then the wall panels.
- Fit the panels easily into place without force, maintaining **moderate contact** between them.
- Always match edges with edges and ends with ends. For example, put tapered end adjacent to tapered end, and match square-cut end to square-cut end.
- Plan to span the entire length of a ceiling or wall with single boards, if possible, to reduce the number of end joints, which are more difficult to finish than edge joints.
- Stagger end joints and locate them as far from the center of the ceiling or wall as possible so they will be inconspicuous.
- In a single-ply application, the board ends or edges parallel to the framing members should fall on these members to reinforce the joint. Do not locate joints in the same place on two sides of a wall assembly. When a joint occurs on a stud on one side of the assembly, do not also locate a joint on the same stud on the other side of the assembly.
- Mechanical and electrical equipment, such as cover plates, registers, and grilles, should be installed to provide for the final wall thickness when applying the trim.
- Keep gypsum boards clean and dry throughout the installation to facilitate the finishing process.

Applying Gypsum Board to Wood Stud Framing with Nails or Screws

Before you install gypsum board on wood stud framing or ceiling joists, carefully inspect the framing itself. Acceptable wood framing consists of properly cured, straight, correctly spaced lumber. Framing faults prevent solid contact between the gypsum board and the framing members, weakening the assembly structure. Moist, misaligned, or twisted supporting framing members and improperly installed blocking or bracing are common causes of face paper fractures when installing gypsum board on wood stud assemblies, as shown in *Figure 21*.

The moisture content of wood framing members should not exceed 15 percent at the time of gypsum board application. Because lumber shrinks across the grain as it dries, applying panels to wood studs with a higher moisture content

Moderate contact: The contact between the edges and ends of abutting gypsum boards in a wall or ceiling assembly, which should not be tight or too widely spaced.

NOTE

The depth of electrical boxes in a wall assembly should not exceed the framing depth, and boxes should not be placed back-to-back on opposite sides of the same stud. Electrical boxes and other devices should not be allowed to penetrate completely through the walls. This is detrimental to both sound control and fire resistance. Before installing gypsum board, make sure all the wires in the electrical boxes are tucked as far back as possible into the boxes to avoid nicking them when using a drywall router to cut out the gypsum board around the box.

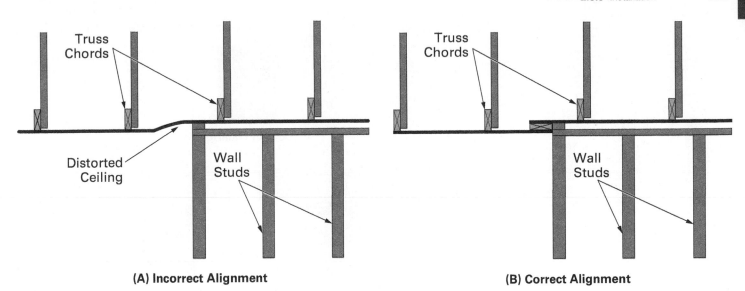

Figure 21 Incorrect and correct alignment.

will eventually expose the shanks of nails driven into the edges of the framing members. If shrinkage is substantial or the nails are too long, separation between the gypsum board and its framing member can result in fastener pops and loose boards. Also, a framing member that sticks out farther than the framing members on either side of it will result in a wall that is not even. No framing member should protrude more than $1/8$" farther than the adjacent members.

If you notice defective supports, you must correct them prior to the application of the gypsum board. Wood framing with a high moisture content should be allowed to dry further in appropriate environmental conditions. Protruding framing members should be trimmed or reinstalled. If fixing the framing or reframing is not possible, you can use wood shims to create a level plane for gypsum board application. If other options are unavailable, two-ply construction can also minimize framing defects.

When you are applying gypsum board to wood ceiling joists, some project plans will call for you to install wood cross furring perpendicular to the joists to provide better support for the ceiling panels.

For single-ply wall installation, when you are applying gypsum boards horizontally (perpendicular) to wood framing members, place all the ends of the boards on the members or other solid backing. When applying gypsum boards vertically, place all board edges on framing members.

You can apply gypsum board to wood studs and joists with drywall nails or drywall screws. The length of the fastener depends on the thickness of the board. See *Table 3* to determine appropriate nail length when using annular or smooth shank nails. The nail penetration into the framing member should be at least $7/8$" for smooth shank nails and $3/4$" for annular nails.

TABLE 3 Nail Length for Single-Ply Standard Gypsum Board Application on Wood Framing or Furring

Gypsum Board Thickness	Minimum Nail Length, Annular Nail	Minimum Nail Length, Smooth Shank Nail
$1/4$ inch (6.4 mm)	Only used in multi-ply assemblies	Only used in multi-ply assemblies
$3/8$ inch (9.5 mm)	$1\,1/8$ inch (28 mm)	$1\,1/4$ inch (32 mm)
$1/2$ inch (12.7 mm)	$1\,1/4$ inch (32 mm)	$1\,3/8$ inch (35 mm)
$5/8$ inch (15.9 mm)	$1\,3/8$ inch (35 mm)	$1\,1/2$ inch (38 mm)

Hold each gypsum board firmly against the framing members and drive each nail perpendicular to the board. This method will help you avoid **fastener pops**, which are protrusions of nail or screw heads above the surface of the gypsum board. The results of correct and incorrect nailing are shown in *Figure 22*.

Fastener pops: Protrusions of nails or screws above the surface of a gypsum board, usually caused by shrinkage of wood framing or by incorrect board installation.

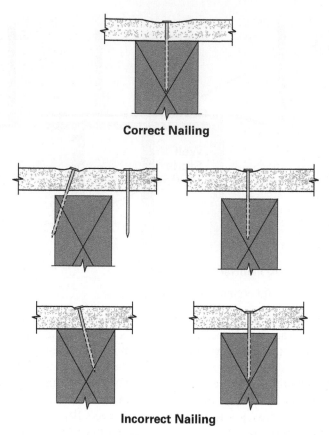

Figure 22 Correct and incorrect nailing.

Perimeter: The outer boundary of a gypsum panel.

Floating angle method: A drywall installation technique used with wood stud framing in which no fasteners are used where ceiling and wall panels intersect in order to allow for structural stresses.

Begin fastening in the inner area of the board, the field, starting in the middle and proceeding outward toward the board edges and ends, the **perimeter**. Around the perimeter of each board, place fasteners at least $^3/_8$" and not more than 1" from the edges and ends—except where the **floating angle method** is appropriate. Use the appropriate hammer to countersink each nail without breaking the face paper of the board.

In single-ply construction, space nails a maximum of 7" on center on ceilings and 8" on center on walls. See *Figure 23*. In multi-ply construction or when installing a single layer of board over rigid foam insulation, make sure to use nails that are long enough to penetrate the wood framing members at least $^7/_8$".

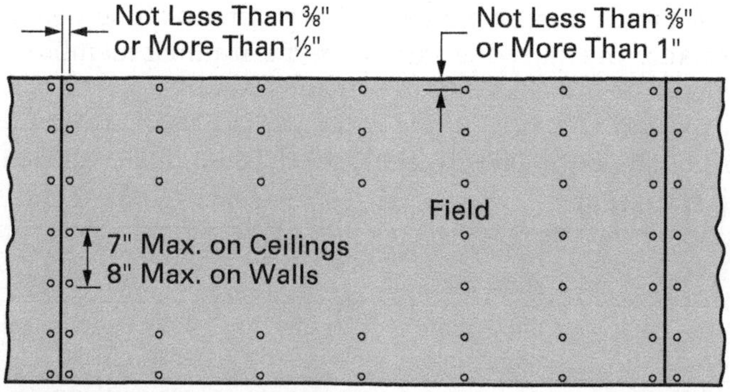

Note:
If screws are used in place of nails, spacing is as follows:

Framing Spacing	Walls	Ceilings
16" OC	16"	12"
24" OC	12"	12"

Figure 23 Single nail spacing.

Drywall screws are more commonly used than nails in gypsum board installation. While drywall nails are less expensive, screws provide a stronger hold, so you use fewer of them. Modern drywall screw guns come with an adjustable depth setting, which makes it relatively easy to consistently countersink your screws. These characteristics make screws and drywall screw guns more popular than hammers and nails among workers who install gypsum panels. Additionally, you can only use drywall nails to apply gypsum board to wood stud assemblies. In comparison, you can use drywall screws to apply gypsum board to both wood and steel stud assemblies.

When using screws to apply gypsum panels to wood stud framing, use #6 drywall screws long enough to penetrate the framing members at least $\frac{5}{8}$". See *Table 4*.

TABLE 4 Screw Length for Single-Ply Standard Gypsum Board Application on Wood Framing or Furring

Gypsum Board Thickness	Drywall Screw, Minimum Length
$\frac{1}{4}$" (6.4 mm)	Only used in multi-ply assemblies
$\frac{3}{8}$" (9.5 mm)	1" (25 mm)
$\frac{1}{2}$" (12.7 mm)	$1\frac{1}{8}$" (28 mm)
$\frac{5}{8}$" (15.9 mm)	$1\frac{1}{4}$" (35 mm)

These professional recommendations are important to consider, but it is also crucial to know that gypsum board installers operate somewhat differently in the field, where the standard screw used to apply gypsum board to wood or steel framing is a #6 drywall screw $1\frac{1}{4}$" long.

The spacing of screws in a single-ply gypsum board assembly on wood stud framing depends on the spacing of the framing members. When wall or ceiling framing members are 16" apart on center, space screws 12" apart on center for ceilings and 16" apart on center for walls. When the wall or ceiling framing members are 24" apart on center, space screws not more than 12" apart on centers for both walls and ceilings. For such widely spaced framing members, double screws are recommended.

When using screws—as when using nails—hold the gypsum board tightly against the framing members while applying the fasteners to avoid fastener pop. When installing multi-layer gypsum board or single-layer board over rigid foam insulation, make sure to use screws long enough to penetrate the wood framing members at least $\frac{5}{8}$".

Floating Angle Method

When you are installing one layer of gypsum panels to wood stud framing, it is necessary to take preventive measures to minimize the possibility of fastener pop in areas adjacent to wall and ceiling intersections and to minimize cracking in gypsum boards due to structural stresses. The floating angle method, which is used only with wood stud framing, omits fasteners where walls and ceilings intersect. This method may be used for either single-ply or multi-ply gypsum board wall assemblies with wood framing, and it is applicable whether you are using nails or screws. *Figure 24* shows a typical single-layer application.

Where the ceiling framing members are perpendicular to the wall/ceiling intersection, the ceiling fasteners should be located 7" from the intersection for single nailing and 11" to 12" for screw applications. Where ceiling joists are parallel to the wall/ceiling intersection, nailing should start at the intersection. See *Figure 25*.

Applying Gypsum Board to Wood Stud Framing with Adhesive

Drywall adhesives should be applied with a caulking gun in accordance with the manufacturer's recommendations. When you are using adhesive to make a joint between gypsum panels on a stud, apply two parallel straight beads of adhesive approximately $\frac{1}{4}$" in diameter, one near each edge of the stud. When you are applying the field of the panel, rather than an end or edge, to a stud, apply one straight

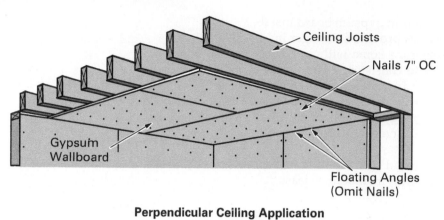

Ceiling Joists

Nails 7" OC

Gypsum Wallboard

Floating Angles (Omit Nails)

Perpendicular Ceiling Application (Single Nailing)

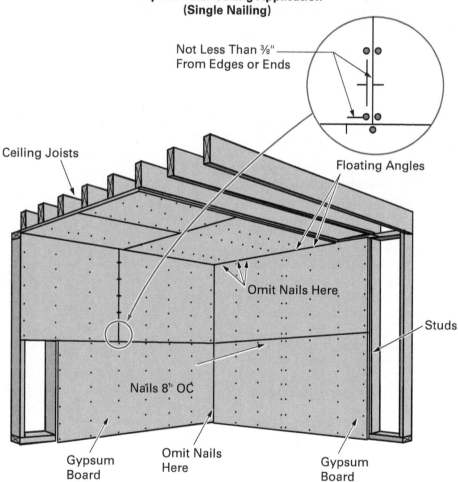

Not Less Than ³⁄₈" From Edges or Ends

Floating Angles

Ceiling Joists

Omit Nails Here

Studs

Nails 8" OC

Gypsum Board

Omit Nails Here

Gypsum Board

Parallel Ceiling Application (Single Nailing)

Figure 24 Floating angle method.

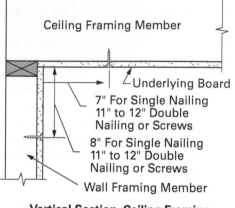

Ceiling Framing Member

Underlying Board

7" For Single Nailing 11" to 12" Double Nailing or Screws

8" For Single Nailing 11" to 12" Double Nailing or Screws

Wall Framing Member

Vertical Section, Ceiling Framing Perpendicular to Wall

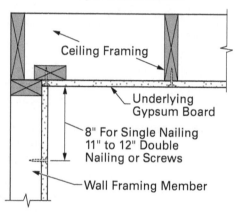

Ceiling Framing

Underlying Gypsum Board

8" For Single Nailing 11" to 12" Double Nailing or Screws

Wall Framing Member

Vertical Section, Ceiling Framing Parallel to Wall

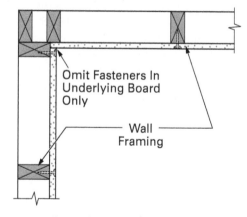

Omit Fasteners In Underlying Board Only

Wall Framing

Cross Section Through Interior Vertical Angle

Figure 25 Fastener patterns for the floating angle method.

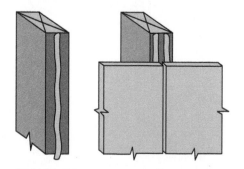

Figure 26 Adhesive applied to the edges of a stud.

bead of adhesive, approximately ¹⁄₄" in diameter, to the middle of the stud face. See *Figure 26*.

Do not expect adhesives alone to hold gypsum boards to wood framing members. Single-ply gypsum board systems attached with adhesives require supplemental perimeter fasteners. Place fasteners 16" on center along the edges or ends on boards that you have applied with adhesive, whether they are perpendicular or parallel to the supports. Ceiling installations require supplemental fasteners in the field as well as on the perimeter. They should be placed 24" on center. Follow the floating angle method for supplemental fasteners used with adhesive. See *Figure 27*.

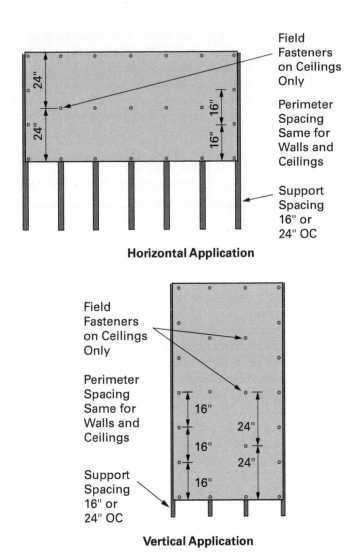

Figure 27 Supplemental wall and ceiling fasteners.

Adhesive is not required at top or bottom plates, bridging, bracing, or fire stops. Where fasteners at vertical joints are undesirable, gypsum panels may be prebowed, as shown in *Figure 28*.

Using Adhesive on Drywall

When applying gypsum board with adhesive, choose a product that meets the requirements of *ASTM C557*. Allow the adhesive to dry for 48 hours before finishing the joints. Using adhesive does not eliminate the need for fasteners; it simply reduces the number of fasteners needed. Check the job specifications or local codes for details.

Adhesives cannot be used to attach drywall panels to studs if the building has an inside moisture barrier.

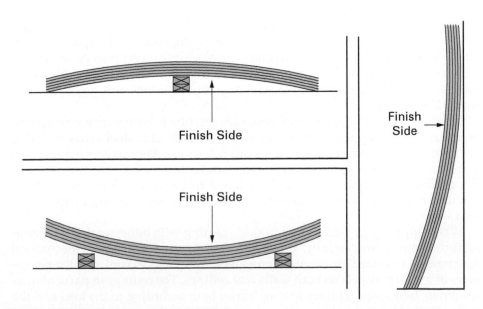

Figure 28 Prebowing of gypsum panels.

Prebowing puts an arc in the gypsum board, which keeps it in tight contact with the adhesive after the board is applied. Supplemental fasteners (placed 16" on center) are then used at the top and bottom plates.

Gypsum board may be prebowed by stacking it, face up, with the ends resting on 2' × 4' lumber or other blocks, and with the center of the boards resting on the floor. Allow it to remain overnight or until the boards have a permanent bow.

Applying Gypsum Board to Steel Stud Framing

For single-ply installation on steel stud framing, attach gypsum board to 18 mil (25 gauge) or 27 mil (22 gauge) steel framing and furring using #6, fine, self-tapping drywall screws. If the steel framing is 30 mil (20 gauge) or thicker, #6, fine, self-drilling drywall screws are required.

The recommended minimum screw penetration into steel studs for single-ply construction is $3/8$" or 3 threads. This length is less than the penetration required for wood studs because the hold between drywall screws and steel framing creates more withdrawal resistance than the hold between screws and wood framing. Keep in mind that these guidelines refer to the length of the threaded part of the screw. Self-drilling screws have a longer tip than other screws, and it is a good idea to allow an additional $1/4$" when choosing these screws to get the recommended penetration into the framing members.

Gypsum board installers applying gypsum board to steel studs most commonly use #6, fine, self-tapping or #6, fine, self-drilling screws that are $1\frac{1}{4}$" long for single-ply construction and $1\frac{5}{8}$" long for multi-ply construction. Choose the length of the screws you use for a particular project based on the guidelines in *Table 5*, by the length designated in the project specifications, or by the length commonly used in the field.

TABLE 5 Screw Length for Single-Ply Standard Gypsum Board Application on Steel Framing or Furring

Gypsum Board Thickness	Minimum Screw Length, Single-Ply Construction
$1/4$" (6.4 mm)	Only used in multi-ply assemblies
$3/8$" (9.5 mm)	$3/4$" (19 mm)
$1/2$" (12.7 mm)	$7/8$" (22 mm)
$5/8$" (15.9 mm)	1" (25 mm)

As with wood stud framing, the spacing of the screws when installing single-ply gypsum panels to steel stud framing depends on the spacing of the studs: When the framing is 16" on center, space screws on ceilings no more than 12" on center and on walls no more than 16" on center. When the framing of the steel studs is 24" on center, space screws on both ceilings and walls no more than 12" on center.

Most ceiling framing in commercial construction is built with a grid system of main beams supported by wires and perpendicular steel cross tees that click into place. These ceiling systems provide an excellent structure for applying gypsum panels, because the boards can be fastened all the way around the perimeter, compared to boards fastened to parallel ceiling joists, which, without installing furring, can only be fastened to the framing members on two sides.

When applying gypsum board to steel framing with adhesive, follow manufacturer's directions carefully. You must supplement adhesive with mechanical fasteners on intermediate beams or cross tees on ceilings as well as at the perimeters of gypsum panels on both walls and ceilings. For ceilings in particular, as shown in *Table 6*, the framing spacing varies both according to the load and the type of board being used.

TABLE 6 Maximum Spacing of Ceiling Framing

Gypsum Board (Thickness)		Application to Framing		Maximum OC Spacing of Framing
Base	Face	Base	Face	
$3/8$"*	$3/8$"	Perpendicular	Perpendicular or Parallel	16"
$1/2$"*	$3/8$" or $1/2$"	Perpendicular or Parallel	Perpendicular or Parallel	16"
$5/8$"*	$1/2$"	Perpendicular or Parallel	Perpendicular or Parallel	16"
$5/8$"*	$5/8$"	Perpendicular or Parallel	Perpendicular or Parallel	24"

*In two-ply construction, the threaded portion of gypsum drywall screws must penetrate at least $1/2$" into the base layer board. To ensure this minimum penetration, it is wise to allow approximately $1/4$" for the screw point by choosing a screw $3/4$" longer than the width of the gypsum board you are installing.

2.1.5 Resurfacing Existing Construction

Gypsum board may be used to provide a new finish on existing above-grade walls and ceilings of wood, plaster, wallboard, masonry, or concrete. If the existing surface is structurally sound and provides a sufficiently smooth and solid backing without shimming, $1/4$" gypsum board can be applied directly to the surface with nails, screws, or adhesives. If you are using drywall nails to apply gypsum panels to existing paneling or plaster, make sure the nails penetrate the framing beneath the old surface material by $7/8$". When using power-driven screws, make sure the threaded portion of the screw penetrates the framing by at least $5/8$".

Applying Gypsum Board Directly to Concrete and Masonry

Before beginning gypsum board installation, remove and set aside any surface covers for mechanical and electrical equipment, such as switch plates, outlet covers, and ventilating grilles. Electrical boxes should be reset prior to the installation of new gypsum board. Then address any irregularities on the surface of the wall. Rough or protruding edges and excess joint mortar should be removed, and any depressions filled with mortar to make the wall surface smooth and level. Base surfaces should also be cleaned of any curing compound, loose particles, dust, and grease to ensure an adequate bond. New concrete should be allowed to cure for at least 28 days before gypsum board is adhered directly to it.

Another acceptable way to create a flat plane for applying gypsum boards is to **fur out** the concrete or masonry surface with wood furring strips or steel furring channels. Furring also provides a separation between existing exterior walls and gypsum panels to help manage possible moisture issues. When applying gypsum panels to wood furring, use fasteners of a length that will penetrate the wood without coming into contact with the concrete or masonry surface beneath it.

When installing $1/4$" gypsum board directly to concrete or masonry with adhesive, choose an adhesive recommended by the manufacturer and follow all manufacturer's instructions and precautions carefully. In addition, use supplemental mechanical fasteners spaced 16" on center to hold the gypsum board in place while the adhesive is developing a bond.

Gypsum board should not be installed directly on below-grade exterior walls. If a project requires you to install gypsum board in such a setting, you should first fur out the walls and apply a vapor barrier and insulation before attaching the panels. Gypsum board can also be adhered directly to exterior cavity walls if the cavities are properly insulated to prevent condensation and the inside face of the cavity is properly waterproofed.

Fur out: To attach furring strips or furring channels to masonry walls, wood framing, or steel framing to create a level surface before applying gypsum board.

NOTE

An alternative to anchoring furring to a masonry wall is to build a $1\frac{5}{8}$" metal stud wall in front of the masonry wall for the application of gypsum board.

2.2.0 Drywall Trim

When you have completed the process of applying gypsum panels to a wall assembly, all edges of wallboard exposed to view will need drywall trim. Adding drywall trim—a long, narrow strip of metal, vinyl, or paper tape—to these joints creates the illusion of a smooth unbroken surface once the drywall has been finished. Trim comes in a variety of shapes, lengths, and widths, each one having a particular function. It can be made of metal (often galvanized metal) or vinyl (see *Figure 29*).

Figure 29 Drywall trim.
Source: Gintare Stackunaite/Shutterstock

Corner bead: A metal or plastic angle used to protect and finish outside corners where drywall panels meet.

Casing: A type of drywall trim that is used around windows and doors.

Flange: The rim of an accessory used to attach it to another object or surface.

Corner bead is a type of drywall trim made to mask the joints at outside corners of a room. Corner bead protects and reinforces outside corner joints, which are prone to wear and tear. It also provides a straight guide for finishing. **Casing** is a type of drywall trim used around windows and doors. Choose corner bead or casing with a wider **flange** if you need to hide flaws in problem areas in a drywall installation.

Corner bead or casing is relatively easy to apply with mechanical fasteners, drywall compound, or adhesive. Some metal corner bead comes with predrilled holes for mechanical fasteners or a perforated or mesh flange that can be effectively embedded in joint compound during application for a strong hold. The disadvantage of using metal corner bead is that if it is not handled carefully before installation, it can be easily damaged. Vinyl corner bead is less easily damaged, and, like metal bead, it can be applied with mechanical fasteners, drywall compound, or adhesive. Some vinyl corner bead, however, requires special compound or adhesive. Follow the manufacturer's directions carefully when applying any type of trim.

Metal and vinyl drywall trim also come in paper-faced options. This paper face, laminated over metal or vinyl bead or casing, blends in well with the paper face of the gypsum board, is easier to finish, and is more durable than trim covered with corner bead and joint compound alone. United States Gypsum (USG) makes a metal paper-faced corner bead, and No-Coat makes a vinyl paper-faced bead.

Some joints between gypsum boards are especially challenging to finish, such as arches and splayed angles. For these difficult areas, consider using paper tape fortified with plastic mesh or metal strips. This flexible tape can be applied using joint compound to inside and outside corners as well as in places that feature decorative elements where extra flexibility is needed.

Casings come in different shapes made for use in specific locations, usually where gypsum board meets other materials, such as a wooden window jamb or a fireplace. The cross-section of L-bead casing is shaped like the letter L, with one flange wider than the other. Use L-bead to cover the joints around window and door jambs. The cross-section of J-bead casing (sometimes called U-bead) is shaped like the letter J. Use J-bead to cap the end of a piece of gypsum board before applying it to the wall assembly where the wall meets another type of surface, such as a countertop or fireplace.

Reveal trim is a special insert that fits into a deliberate space left between gypsum boards to create a decorative effect in a wall or ceiling assembly. Reveals might be used as moldings around room perimeters, door frames, windows, archways, and other architectural features. They might also be applied at regular intervals in a wall or ceiling assembly to create a decorative pattern.

Some reveal trim must be installed at the time the gypsum panels are hung. If reveal trim is part of your drywall installation project, follow the project plans carefully. Drywall reveal is available in different shapes designed for different purposes. A standard drywall reveal, for example, is meant to be installed between two gypsum panels. It features a U-shaped main channel with two tapered flanges on either side. When the standard reveal is placed between the two boards, the flanges sit on top of the boards and are fastened to them with drywall fasteners through predrilled holes in the flanges.

In comparison, drywall F reveals have one tapered flange and are used to create a recessed molding around the perimeter of a room, where the wall meets the ceiling. The top of the U-shaped main channel of an F reveal lies flush against the gypsum board on the ceiling, and the flange is attached to the gypsum board on the wall below the reveal with a drywall fastener. F reveals are finished with joint tape and joint compound.

When you are ready to begin installing drywall trim, measure the length of the surface carefully. Then use aviation snips to cut the trim to fit.

Achieving a Rounded Appearance

A smooth, rounded finish appearance can be obtained by using a bullnose corner molding and cap such as the ones shown here.

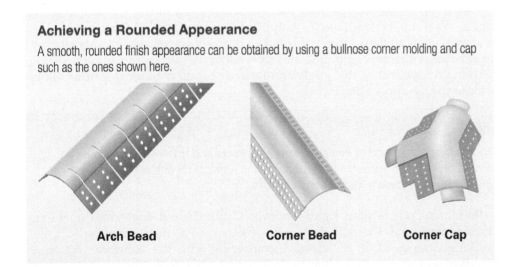

Arch Bead   **Corner Bead**   **Corner Cap**

2.3.0 Moisture Resistance

When a project calls for gypsum board installation in areas such as in laundry rooms, kitchens, bathrooms, and basements, give special consideration to the appropriate use of gypsum panels. In certain wet or humid settings, standard or moisture-resistant gypsum board might be appropriate, while in others its use is not recommended. At times, project managers and consumers must decide whether to follow manufacturer's product recommendations, professional standards, or local building codes. Sometimes local building codes are not as stringent as professional standards.

2.3.1 Moisture-Resistant Construction

Many brands of gypsum board offer moisture-resistant panels, and some local building codes allow these panels to be used as a base for applying tile with adhesive in wet areas, such as tubs, showers, and saunas. However, ASTM specifications (*ASTM C840*) recommend using both standard and moisture-resistant gypsum board only in dry areas where tile will be applied. In wet settings where a base is needed for tile application, cement backer board is the better option. Leading gypsum board manufacturer USG does not recommend using its gypsum board products in wet settings but instead offers its own water-durable cement backer board.

If the project you are working on does require you to use gypsum board in moist conditions, do not use foil-backed board or apply board directly over a vapor barrier, because the vapor barrier will trap moisture within the gypsum core of the board. In areas of high humidity or where water vapor is present, such as in basements, exterior below-grade walls or surfaces should be furred out and protected with a vapor barrier and insulation in order to provide a suitable base for attaching the moisture-resistant gypsum board.

If project plans call for you to install moisture-resistant gypsum board in a tub/shower enclosure as a backer for tile, make sure the shower pan or tub has been installed before you begin. Shower pans should have an upstanding lip or flange located at a minimum of 1" higher than the entry wall to the shower. It is recommended that the tub be supported. If necessary, fur out the framing members before applying gypsum board so the upstanding leg of the pan (*Figure 30*) will be on the same plane as the face of the board.

Suitable blocking should be provided approximately 1" above the top of the tub or pan. Between-stud blocking should be placed behind the horizontal joint of the board above the tub or shower pan. For ceramic tile applications, use studs that are at least $3\frac{1}{2}$" deep and placed 16" on center. Appropriate blocking, headers, or supports should be provided for tub plumbing fixtures and to receive soap dishes, grab bars, towel racks, and similar items.

Install tile backer boards with nails or screws spaced not more than 8" on center. When the ceramic tile to be applied is more than $\frac{3}{8}$" thick, space nails or screws no more than 4" on center. When it is necessary to treat joints and fastener heads with joint compound and tape, either use waterproof nonhardening caulking compound or seal joints and fastener heads with a compatible sealer prior to the tile application.

NOTE

Different types of waterproof boards have different applications and limitations, so it is always necessary to check the manufacturer's product data sheets, installation instructions, and local building codes for the type of board to be used.

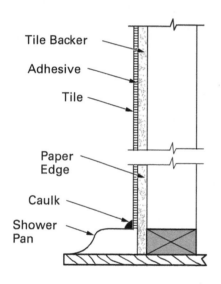

Figure 30 Pan is on the same plane as the face of the board.

NOTE

The caulking compound or sealer used to seal joints and fastener heads must be compatible with the adhesive to be used for the application of the tile. Follow the adhesive manufacturer's instructions carefully.

Reinforce corners with rigid supports. Caulk the cut edges and openings around pipes and fixtures flush with a waterproof, nonhardening, silicone caulking compound or an adhesive complying with the American National Standard for Organic Adhesives for Installation of Ceramic Tile.

Tile Application

Ceramic wall tile application to gypsum board should meet the American National Standard Specifications for Installation of Ceramic Tile with Water-Resistant Organic Adhesive. The adhesives used should meet the American National Standard for Organic Adhesives for Installation of Ceramic Tile.

The surfacing material should be applied down to the top surface or edge of the finished shower floor, return, or tub, and installed to overlap the top lip of the receptor, subpan, or tub.

2.4.0 Fire-Resistance-Rated and Sound-Rated Assemblies

The construction of ceilings, exterior walls, and interior partition walls in certain settings is driven by the fire resistance and acoustical performance requirements specified in local building codes. In buildings in which fire and sound rating are not required—such as in single-family residential homes—wood or steel stud frame wall and ceiling assemblies with one layer of $\frac{1}{2}$" gypsum drywall are satisfactory. In contrast, multi-ply ceiling and wall assemblies are necessary in construction that requires a one- or two-hour fire-resistance rating. For an even higher fire-resistance rating, such as in offices located next to the manufacturing spaces in a factory where there is a hazard of fire or explosion, it may be necessary to build multi-ply ceiling and wall assemblies that begin with a concrete block (CMU) wall, followed by rigid and/or fiberglass insulation, followed by fire-resistant gypsum board, as shown in *Figure 31*.

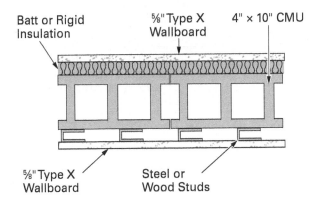

Figure 31 High fire/noise resistance partition.

Wood and steel stud framing are both approved for use in ceiling and wall assemblies in single-family residential construction, but steel stud framing is the standard in multi-family residential and commercial construction, where fire-resistance and sound rating are required by building codes. The type and thickness of the insulation and gypsum board applied to the steel stud framing depend on the fire-resistance rating and acoustical performance requirements for the project. Multi-family residential construction is required to be one-hour fire-resistance rated. Sound control requirements are based on the amount of privacy required for the building's intended use. For example, high-rise condos, executive offices, hospitals, and homes for the elderly may require more privacy than general offices.

The requirements for sound attenuation and fire resistance can significantly affect the thickness of a wall. For example, a steel stud wall in an assembly with high sound and fire-resistance ratings might have a total thickness of nearly $6\frac{1}{2}$", while a wall in an assembly with a lower rating might have a total thickness of only $3\frac{1}{2}$".

2.4.1 Fire-Resistance-Rated Construction

Building codes for certain types of structures, such as multistory apartment and office buildings, require specific wall and ceiling assemblies. A **fire-resistance-rated assembly** can withstand fire for a certain period, measured in hours, as determined under laboratory conditions. A fire-resistance-rated assembly can effectively keep fire from spreading while maintaining the integrity of the assembly for a specified length of time. A single-family residential home may use a **non-rated assembly**, which may provide some resistance to fire, but not enough to qualify for any fire-resistance rating.

Fire-resistance-rated assembly: Construction built with certain materials in a certain configuration that has been shown through testing to restrict the spread of fire.

Non-rated assembly: A ceiling or wall assembly that does not exhibit enough fire-resistant properties to qualify for a fire-resistance rating.

There are many different construction methods for so-called party walls, or walls common to two adjoining buildings or rooms. Each is designed to meet different fire and sound control standards. The wall is generally at least 3" thick and contains several layers of gypsum board and insulation. A fire-resistance-rated wall may abut a non-rated partition or wall. Ensure that the rated wall is carried through to maintain the fire-resistance rating, as shown in *Figure 32*.

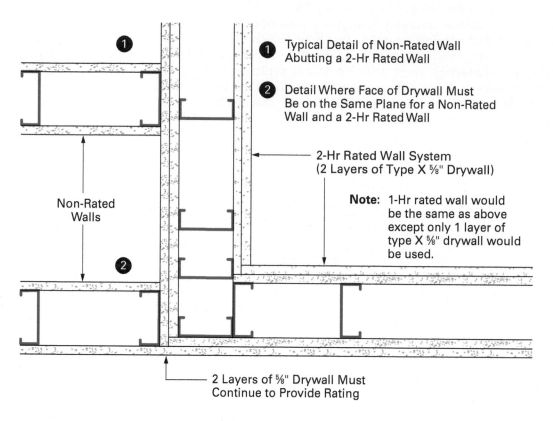

① Typical Detail of Non-Rated Wall Abutting a 2-Hr Rated Wall

② Detail Where Face of Drywall Must Be on the Same Plane for a Non-Rated Wall and a 2-Hr Rated Wall

2-Hr Rated Wall System (2 Layers of Type X ⅝" Drywall)

Note: 1-Hr rated wall would be the same as above except only 1 layer of type X ⅝" drywall would be used.

Non-Rated Walls

2 Layers of ⅝" Drywall Must Continue to Provide Rating

Figure 32 An example of a fire-resistance-rated wall abutting a non-rated wall.

Fire-Resistance-Rate Gypsum Board

Type X gypsum board is the most common type used in fire-resistance-rated assemblies. It contains noncombustible glass fibers in its gypsum core, which keep the board from shrinking during the calcination that happens during a fire.

Type C gypsum board is an enhanced, more expensive version of gypsum board that is also approved for fire-resistance-rated assemblies. In addition to glass fiber, it contains unexpanded vermiculite in its gypsum core, which expands when heated to keep the board from shrinking and causing a gap in the fire barrier.

Firestopping

Firestopping means cutting off the air supply so that fire and smoke cannot readily move from one location to another.

In frame construction, a firestop is a piece of wood or fire-resistant material inserted into an opening such as the space between studs. This firestop acts as a barrier to block airflow that would feed and carry a fire to the upper floors.

In commercial construction, firestopping material is used to close wall penetrations such as those created to run conduit, piping, and air conditioning ducts. If such openings are not sealed, fire will travel through the openings.

Avoid Back-to-Back Fixtures

Medicine cabinets; electrical, telephone, television, and intercom outlets; and plumbing, heating, and air conditioning ducts should not be installed back-to-back. Any opening for such fixtures, piping, and electrical outlets should be carefully cut to the proper size and caulked.

To meet the fire-resistance-rating standards established by the building and fire codes, all openings must be sealed. The firestopping methods used for this purpose are classified as mechanical and nonmechanical.

Mechanical firestops are fire-resistant devices used to seal the space around wiring or piping. There are many different options, such as lighting covers, sleeve kits, cableways, cable transits, pathways, grommets, plugs, and collars.

Nonmechanical firestops are fire-resistant materials, such as caulks and putties, that are used to fill the space around the conduit or piping. You may be required to install various nonmechanical firestopping materials when working with fire-resistance-rated walls and floors. Holes or gaps affect the rating of a floor or wall. Properly filling these penetrations with firestopping materials maintains the rating. Firestopping materials are typically applied around all types of piping, electrical conduit, ductwork, electrical and communication cables, and similar devices that run through openings in floor slabs, walls, and other fire-resistance-rated building partitions and assemblies.

Nonmechanical firestopping materials are classified as intumescent or endothermic. Both are formulated to help control the spread of fire before, during, and after exposure to open flames. When subjected to the extreme heat of a fire, intumescent materials expand (typically up to three times their original size) to form a strong insulating material that seals the opening for three to four hours. Should the insulation on the cables, pipes, etc., passing through the penetration become consumed by the fire, the expansion of the firestopping material also acts to fill the void in the floor or wall to help stop the spread of smoke and other toxic products of combustion.

Did You Know?

Firestopping Versus Fireproofing

Firestopping and fireproofing are not the same thing. Firestopping is intended to prevent the spread of fire and smoke from room to room through openings in walls and floors. Fireproofing is a thermal barrier that causes a fire to burn more slowly and retards the spread of fire.

Endothermic materials block heat by releasing chemically bound water, which causes them to absorb heat.

Firestopping materials are formulated in such a way that when activated, they are free of corrosive gases, reducing the risks to building occupants and sensitive equipment.

Firestopping materials are made in a variety of forms, including composite sheeting, caulks, silicone sealants, foams, moldable putty, wrap strips, and spray coatings. They come in both one-part and two-part formulations. The installation of these materials must always be done in accordance with the applicable building codes and the manufacturer's instructions for the product being used. Depending on the product, firestopping materials can be applied via spray equipment, conventional caulking guns, pneumatic pumping equipment, or a putty knife.

2.4.2 Sound-Rated Construction

The first step for sound isolation of any assembly is to close off air leaks and flanking paths. A flanking path is when sound leaks around or through building components. Noise can travel through the air over, under, or around walls; through the windows and doors; through air ducts; and through floors and crawl spaces. All these paths must be correctly treated.

Sound-rated assembly: Construction built with certain materials in a certain configuration that has been shown through testing to obtain specific acoustical performance.

Sound transmission class (STC): A rating that describes how much sound a product or assembly prevents from getting through to the other side. The higher the STC, the more sound at common frequencies has been prevented from traveling through a product or assembly to an adjacent space.

Enhanced gypsum board and specialized insulation are available for use in a **sound-rated assembly**. One type, for example, has a core of viscose polymer between two layers of gypsum plaster, a combination that helps to block sound transmission between rooms and dampen sound in large rooms.

Buildings are generally required to meet a **sound transmission class (STC)** rating. The STC is a numeric rating representing the effectiveness of the construction in isolating airborne sound transmission. The higher the STC rating, the better the sound absorption. Hairline cracks and other openings can have an adverse effect on the ability of a building to achieve its STC rating, particularly in higher-rated construction. Where a very high STC performance is needed, air conditioning, heating, and ventilating ducts should not be included in the assembly.

Failure to observe special construction and design details can destroy the effectiveness of the best assembly. Improved sound isolation is obtained by the following:

- Separate framing for the two sides of a wall
- Resilient channel mounting for the gypsum board
- Using sound-absorbing materials in wall cavities
- Using multi-layer gypsum board of varying thicknesses in multi-layer construction
- Caulking the perimeter of gypsum board partitions, openings in walls and ceilings, partition/ mullion intersections, and outlet box openings
- Locating recessed wall fixtures in different stud cavities

The entire perimeter of sound-isolating partitions should be caulked around the gypsum board edges to make it airtight, as detailed in *Figure 33A* and *Figure 33B*. The caulking should be a nonhardening, nonshrinking, nonbleeding, nonstaining, resilient sealant.

Sound-control sealing must be covered in the specifications, understood by all related tradespeople, supervised by the appropriate party, and inspected carefully as the construction progresses.

Separated Partitions

A staggered wood stud gypsum partition placed on separate plates will provide an STC between 40 and 42. The addition of a sound-absorbing material between the studs of one partition side can increase the STC by as much as 8 points. With $\frac{5}{8}$" Type X gypsum board on each side, an assembly has a fire-resistance classification of one hour. Separated walls without framing can also be constructed by using an all-gypsum, double-solid, or semi-solid partition.

Steel or wood tracks fastened to the floor and ceiling hold the partitions in place. For the attachment of kitchen cabinets, lavatories, ceramic tile, medicine cabinets, and other fixtures, a staggered stud wall rather than a resilient wall is recommended. The added weight and fastenings may short circuit the construction acoustically.

Resilient Mountings

Resilient attachments acting as shock absorbers reduce the passage of sound through the wall or ceiling and increase the STC rating. Further STC increases can result from more complex construction methods incorporating multiple layers of gypsum board and building insulation in the wall cavities.

A **resilient furring channel** is attached with the nailing flange down and at a right angle to the wood stud, as shown in *Figure 34*.

Resilient furring channel: A furring channel with one flange to attach to another surface, leaving the other side of the channel unattached.

To install furring channels, drive $1\frac{1}{4}$" coarse screws or 6d coated nails through the prepunched holes in the channel flange. With extremely hard lumber, $\frac{7}{8}$" or 1" fine, self-tapping screws may be used. Locate the channels 24" from the floor, within 6" of the ceiling line, and no more than 24" on center. Extend the channels into all corners and fasten them to the corner framing. Attach $\frac{1}{2}$" × 3" gypsum board filler strips to the bottom plate directly over the studs by overlapping the ends and fastening both flanges to the stud. Apply the gypsum board horizontally with the long dimension parallel to the resilient channels using 1" fine, self-tapping screws spaced 12" on center along the channels. The abutting edges of boards should be centered over the channel flange and securely fastened.

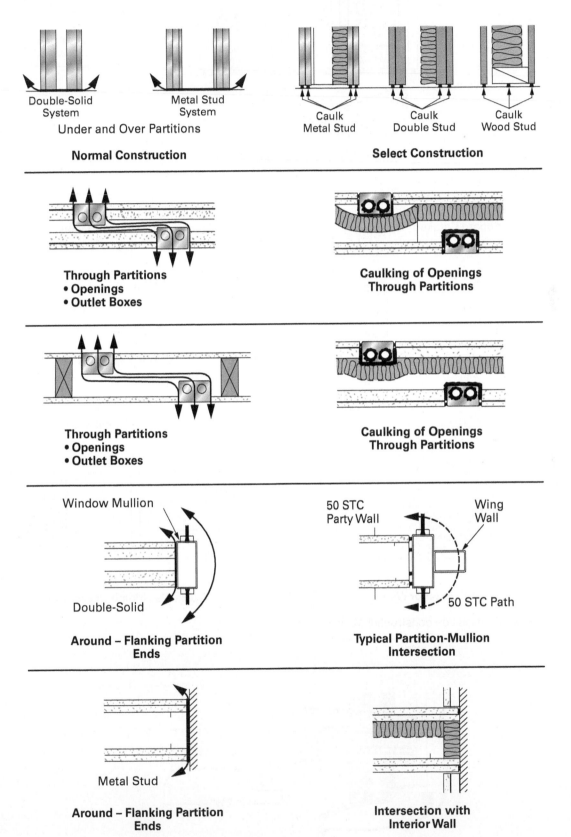

Figure 33A Caulking of sound-isolation construction (1 of 2).

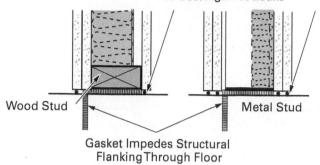

Pre-Design Construction
Simulating Laboratory Conditions

¼" Perimeter Relief and Caulking
to Seal Against Leaks

Wood Stud   Metal Stud

Gasket Impedes Structural
Flanking Through Floor

Typical Floor-Ceiling or Roof Detail

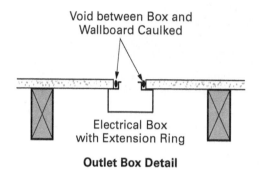

Void between Box and
Wallboard Caulked

Electrical Box
with Extension Ring

Outlet Box Detail

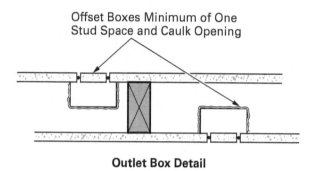

Offset Boxes Minimum of One
Stud Space and Caulk Opening

Outlet Box Detail

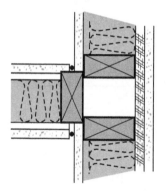

Intersection with Exterior Wall

Wood or Metal Stud

Typical Partition Intersections

Figure 33B Caulking of sound-isolation construction (2 of 2).

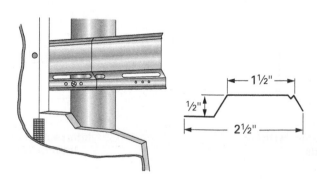

1½"

½"

2½"

Figure 34 Attached resilient furring channel.

Sound-Isolating Materials

Sound-isolating materials include the following:

- Mineral fiber (including glass) blankets and batts used in wood stud assemblies
- Semi-rigid mineral or glass fiber blankets for use with steel studs and laminated gypsum partitions
- Mineral (including glass) fiberboard
- Rigid plastic foam furring systems
- Lead or other special shielding materials

Mineral wool or glass fiber insulating batts and blankets may be used in assembly cavities to absorb airborne sound within the cavity. They should be placed in the cavity and carefully fitted behind electrical outlets and around any cutouts necessary for plumbing lines. Insulating batts and blankets may be faced with paper or another vapor barrier and may have flanges or be of the unfaced friction-fitted type.

Gypsum board may be applied over rigid plastic foam insulation. It is applied on the interior side of exterior masonry and concrete walls to provide a finished wall and to protect the insulation from early exposure to fire originating within the building. Additionally, these systems provide the high insulating values needed for energy conservation.

In new construction or for remodeling, these systems can be installed with as little as 1" dimension from the inside face of the framing or masonry ($\frac{1}{2}$" insulation and $\frac{1}{2}$" Type X gypsum board).

When applying gypsum board over rigid foam insulation, the entire insulated wall surface should be protected with the gypsum board, including the surface above ceilings and in closed, unoccupied spaces.

Single-ply or double-ply, $\frac{1}{2}$" or $\frac{5}{8}$" gypsum board should either be screw-attached to steel wall furring members attached to the masonry or nailed directly into wood framing, as shown in *Figure 35*. Follow the insulation manufacturer's instructions.

Rigid Foam Insulation

Gypsum board applied over rigid plastic foam insulation in the manner described may not necessarily provide the finish ratings required by local building codes. Many building codes require a minimum fire protection for rigid foam on interior surfaces equal to that provided by $\frac{1}{2}$" Type X gypsum board when tested over wood framing. The flammability characteristics of rigid foam insulation products vary widely, and the manufacturer's literature should be reviewed.

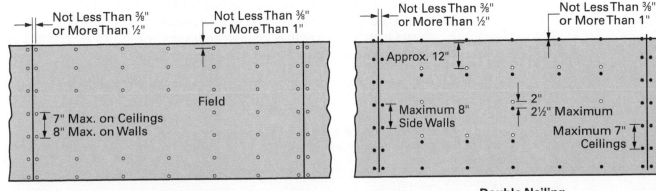

Single Nailing

Double Nailing

Figure 35 Nailing patterns for installation over rigid foam insulation.

Furring members should be designed to minimize thermal transfer through the member and to provide a $1\frac{1}{4}''$ minimum width face or flange for screw application of the gypsum board.

Furring members should be installed vertically and spaced 24" on center. Provide blocking or other backing as required for attachment and support of fixtures and furnishings. Furring members should also be attached at floor/wall and wall/ceiling angles (or at the termination of the gypsum board above suspended ceilings) and around door, window, and other openings. Single-ply gypsum board should be applied vertically, with the long edges of the board located over furring members. The installation should be planned carefully to avoid end joints. The fastener spacing should be the same as required for single-ply application over framing or furring.

In multi-ply applications, the base ply should be applied vertically. In horizontal face ply applications, the face ply and end joints should be offset by at least one framing or furring member space from the base ply edge joints.

The fastener spacing should be as required for multi-ply application over framing or furring, as discussed previously.

In wallboard applications, mechanical fasteners should be of such a length that they do not penetrate completely to the masonry or concrete. In single-layer applications, all joints between gypsum boards should be reinforced with tape. In addition, gypsum board joints should be finished with joint compound. In two-layer applications, the base layer joints may be concealed or left exposed.

2.0.0 Section Review

1. To increase sound control and fire-resistive performance, _____ have two or more layers of gypsum board.
   a. steel framing members
   b. single-ply assemblies
   c. exterior walls
   d. multi-ply assemblies

2. Gypsum board should be kept in the area where it will be installed for at least _____ before installation.
   a. 24 hours
   b. 72 hours
   c. 12 hours
   d. 48 hours

3. A general practice to ensure a sound installation of gypsum board is to _____.
   a. rotate every other piece of gypsum board 180°
   b. start in the middle of a wall and work outward
   c. always match edges with edges and ends with ends
   d. always install wall panels before ceiling boards

4. Corner bead is a type of drywall trim made to mask joints at the _____.
   a. floor
   b. ceiling
   c. inside corner of a room
   d. outside corner of a room

5. In wet settings where a base is needed for tile application, _____ is the better option.
   a. Type X gypsum board
   b. cement backer board
   c. Type C gypsum board
   d. moisture-resistant gypsum board

6. To meet fire-resistance-rating standards for firestopping established by the building and fire codes, _____.
   a. openings must have airflow
   b. mechanical firestop devices must be used on all openings
   c. nonmechanical firestop materials must be used on all openings
   d. all openings must be sealed

Module 45104   Review Questions

1. Gypsum board retards the spread of fire because _____.
   a. its core is exposed on the end
   b. its field and edge are covered with paper
   c. its core is composed of all man-made materials
   d. it contains water in its crystal molecular structure

2. One of the advantages of gypsum board is that it is _____ because it is readily available and easy to apply, repair, decorate, and finish.
   a. economical
   b. sustainable
   c. versatile
   d. durable

3. Gypsum board is considered a(n) _____ material because in addition to naturally mined gypsum, it uses flue gas desulfurization (FGD) gypsum, a synthetic byproduct of coal-burning power plants.
   a. durable
   b. sustainable
   c. versatile
   d. economical

4. If you break the face paper on a gypsum board while driving a nail or screw, _____.
   a. remove the gypsum board
   b. remove the nail or screw
   c. drive the nail or screw farther into the gypsum board so it is invisible
   d. drive another nail or screw properly near the same place

5. When you are installing fire-resistance-rated wallboard, the nails should penetrate the studs by at least _____.
   a. $1/2$"
   b. $3/4$"
   c. $7/8$"
   d. $1\frac{1}{8}$"

6. The head of a drywall hammer has a _____, which transfers to the depression around the nail head and helps joint compound to adhere.
   a. flat surface
   b. waffle texture
   c. curved hook
   d. Phillips head attachment

7. _____ are recommended when attaching gypsum panels to steel framing that is 30 mil (20 gauge).
   a. Self-drilling screws
   b. Self-piercing screws
   c. Annular nails
   d. Coarse screws

8. _____ are designed to fasten gypsum board to wood framing or furring strips.
   a. Fine thread screws
   b. Self-tapping screws
   c. Self-drilling screws
   d. Coarse thread screws

9. When using any type of drywall adhesive, be sure to _____.
   a. apply multiple coats
   b. keep the work area hot to speed up drying time
   c. follow the manufacturer's directions for use
   d. avoid all solvent-based adhesives

10. You would use a _____ to smooth rough-cut edges of gypsum board.
    a. sanding block
    b. rasp
    c. drywall saw
    d. T-square

11. The point at which the edges of two panels of drywall meet is known as a _____.
    a. bedding seam
    b. joint
    c. gypsum lath
    d. depression

12. Gypsum drywall should only be installed with adhesive when the building temperature is greater than _____.
    a. 50°F
    b. 60°F
    c. 70°F
    d. 80°F

13. Fasteners should be applied to wallboard working from _____.
    a. top to bottom
    b. edge to edge
    c. center to edge
    d. corner to corner

14. In floating interior angle construction where the framing is perpendicular to the wall/ceiling intersection, the ceiling fasteners should be located _____ from the intersection.
    a. 1"
    b. 7"
    c. 10"
    d. 12"

15. Drywall trim needs to be applied to all _____.
    a. edges of wallboard exposed to view
    b. multi-ply construction
    c. single-ply construction
    d. wall board that has been secured with adhesive

16. If the project you are working on requires you to use gypsum board in moist conditions, _____.
    a. use multi-ply construction
    b. do not apply board directly over a vapor barrier
    c. use drywall contact adhesive to secure the board
    d. use parallel construction

17. _____ ceiling and wall assemblies are necessary in construction that requires a one- or two-hour fire-resistance rating.
    a. Wood framed
    b. Sound-rated
    c. Single-ply
    d. Multi-ply

18. When subjected to the extreme heat of fire, _____ materials expand to form a strong insulating material that seals an opening for three to four hours.
    a. endothermic
    b. intumescent
    c. moisture resistant
    d. multi-ply

19. Each of the following is a construction method used to control noise except _____.
    a. caulking around outlet box openings
    b. placing air conditioning ducts back-to-back
    c. using separate framing for the two sides of a wall
    d. mounting gypsum board in resilient channels

20. A low STC rating indicates _____.
    a. excellent fire resistance
    b. excellent sound isolation
    c. poor fire resistance
    d. poor acoustical performance

Answers to Odd-Numbered Module Review Questions are found in *Appendix A*.

Answers to Section Review Questions

Answer	Section	Objective
Section One		
1. c	1.1.0	1a
2. a	1.2.2	1b
3. a	1.3.0	1c
4. c	1.4.0	1d
Section Two		
1. d	2.1.1	2a
2. d	2.1.2	2a
3. c	2.1.4	2a
4. d	2.2.0	2b
5. b	2.3.1	2c
6. d	2.4.1	2d

Drywall Finishing

Source: ungvar/Shutterstock

Objectives

Successful completion of this module prepares you to do the following:

1. Identify and describe tools and materials used for drywall finishing, as well as safe practices for using them.
   a. List drywall finishing tools and explain how they are used.
   b. List and describe materials used for drywall finishing.
2. Explain drywall finishing procedures and describe finish levels.
   a. Explain how site conditions affect a drywall project and describe drywall inspection.
   b. Describe the steps of the drywall finishing process.
   c. Describe automatic taping and finishing procedures.
   d. Describe hand finishing procedures.
3. Explain the importance of knowing how to repair damage, list types of damage, and describe steps to fix different kinds of damage.
   a. Describe finished joint repair.
   b. Describe repair procedures for fixing problems with compound.
   c. Describe fastener-related damage and how to repair it.
   d. List problems with wallboard and explain how to fix them.

Performance Tasks

Under supervision, you should be able to do the following:

1. Properly prepare the following compounds for use:
   - Taping compound
   - Topping compound
   - Premix
   - Quick setting compound
2. Select the proper hand tools and perform the following:
   - Joint taping and finishing
   - Fastener spotting
   - Corner finishing
   - Sanding
3. Distinguish a finish level by observation and identify the steps needed to take it to the next level of finish.
4. Patch damaged drywall using two different methods.

Digital Resources for Drywall

Scan this code using the camera on your phone or mobile device to view the digital resources related to this craft.

Overview

Drywall finishing and repair are two important skills for a drywall mechanic. A remodeling project is an example that includes drywall installation and finishing to ensure there are no visible seams between drywall sheets and in corners. Damaged drywall or drywall that was improperly installed must be repaired. Therefore, it is essential to be thoroughly familiar with the tools, materials, and procedures used in drywall finishing and repair.

NCCER Industry-Recognized Credentials

If you are training through an NCCER-accredited sponsor, you may be eligible for credentials from NCCER. The ID number for this module is 45105. Note that this module may have been used in other NCCER curricula and may apply to other level completions. Contact NCCER at 1.888.622.3720 or go to **www.nccer.org** for more information.

You can also show off your industry-recognized credentials online with NCCER's digital credentials. Transform your knowledge, skills, and achievements into credentials that you can share across social media platforms, send to your network, and add to your resume. For more information, visit **www.nccer.org**.

1.0.0 Tools and Materials

Performance Tasks

1. Properly prepare the following compounds for use:
   - Taping compound
   - Topping compound
   - Premix
   - Quick setting compound
2. Select the proper hand tools and perform the following:
   - Joint taping and finishing
   - Fastener spotting
   - Corner finishing
   - Sanding

Objective

Identify and describe tools and materials used for drywall finishing, as well as safe practices for using them.

a. List drywall finishing tools and explain how they are used.
b. List and describe materials used for drywall finishing.

When gypsum drywall is first installed, there are visible seams between the drywall sheets and at corners. These seams must be closed in a way that makes them invisible to anyone looking at the finished wall.

Look around your home. Although the walls were most likely finished with 4' × 8' sheets of drywall, they should look like one solid sheet. The seams are covered with paper or fiberglass **tape**, which is embedded with joint compound, also known as mud. When the joint compound has dried, the joint is sanded flat. This process is usually done three times for each seam. Like many areas of construction, it's not as simple as it sounds. There are many types of compound and a variety of finishing tools used for the drywall finishing process.

Tape: A strong paper or fiberglass tape used to cover the joint between two sheets of drywall.

1.1.0 Finishing Tools

Proper finishing of walls and ceilings would be nearly impossible without a variety of finishing tools. These hand-operated and mechanical tools not only speed up the finishing process, but also help create walls and ceilings that are smooth and flat. This module will introduce you to many tools that are used specifically in finishing procedures.

1.1.1 Tool Safety

Although they appear simple and harmless, finishing tools can be as dangerous as any other kind of tool when handled improperly. You should always follow the manufacturer's recommendations for use and proper PPE. Finishing knives have very sharp edges and corners. Automatic tapers have gears and chains that can catch fingers or hair. Automatic finishers have sharp blades, and some have spring-loaded hatches that can catch fingers, hair, and clothing.

The best way to avoid accidents and tool-related injuries is to treat your tools with respect. Never horse around when tools are being used by you or anyone else. Do not handle any tool unless you have been thoroughly trained in its use. If you do not know how to use a tool, ask your instructor, job supervisor, or another drywall mechanic to teach you.

Always keep your finishing tools clean and free of rust or excess compound, both of which can be poisonous if they get into a cut. Do not use knives with chipped blades or broken handles. Not only can they hurt you, but they can also ruin your finishing work.

Treat automatic tools with special care. They must be cleaned and maintained according to the manufacturer's directions. You can hurt yourself trying to force a mechanical tool to work when the tool is jammed with dried **joint compound**. Inspect automatic tools before every use; if a tool looks damaged or in bad repair, report the problem to your supervisor.

1.1.2 Hand Tools

The following hand tools are used to cut, hang, and finish drywall:

- 4' straightedge or T-square
- Utility knife with plenty of blades
- Drywall saw
- Circle cutter
- Drywall hammer
- Caulking gun
- Screwdriver
- Broad knife
- Joint trowel
- Corner tool
- Mud pan
- Sandpaper/drywall screen
- Sanding block, pole sander, or electric sander
- Sponge sander

An easy way to cut drywall is to place the straightedge on the finished side of the panel and score the drywall with a utility knife, cutting through the paper facing and into the gypsum. Then snap the panel apart along the score line by applying pressure from the back of the board. If the backing paper is still intact, use the knife to finish cutting through it.

An alternative method of cutting through drywall is to use a drywall saw. It is good for making straight as well as curved cuts. A saw with a long, thin, pointed blade (*Figure 1*) works well for limited or detail cuts such as when cutting out a hole for an electrical box. The saw can cut in a straight line or in an arc. Use a drill to make a starter hole for the blade.

Finishing knives range in type and size from the $1\frac{1}{4}$" putty knife (*Figure 2*) to the extremely wide taping knife (*Figure 3*). They are all similar in function, however, because they bed and feather the taping or **topping compound**. Each knife is designed for use in a different situation. For example, the smallest putty knife is used in hard-to-reach spaces, for patching, and for working the compound around windows, cabinets, and doors. The 4", 6", 8", 10", and 12" widths are the most common.

Joint compound: Patching compound used to finish drywall joints, conceal fasteners, and repair irregularities in the drywall. It dries hard and has a strong bond. Sometimes called *mud* or *taping compound*.

Topping compound: A joint compound used for second and third coats. It dries soft and smooth and is easier to sand than taping compound.

Figure 1 Drywall saw with thin, pointed blade for detail cuts.
Source: Stocksnapper/Shutterstock

Figure 2 Putty knife.
Source: cynoclub/Shutterstock

Figure 3 Finishing knives.
Source: Oder01/Shutterstock

Feathering: Tapering joint compound at the edges of a drywall joint to provide a uniform finish.

Taping knives generally have blades made of blue steel, which flexes under the pressure of bedding and **feathering** taped joints. They range in size according to blade width.

A long-handled broad knife is used to wipe excess compound from freshly taped joints. The broad knife's blade is made of stainless steel or blue steel and may range in width from 7" to 9". Typically, a broad knife's handle is about 10" to 12" long but handles up to 28" long are available. These long knives are handy tools for cleaning up messy joints and spatters on high walls and ceilings.

Cornering tools enable you to finish both inside and outside corners. They include not only troweling tools, but also sanding and bead-attaching devices. Corner trowels are made for finishing inside and outside corners. While inside and outside corner trowels are shaped alike, their handles are on opposite sides (*Figure 4*). These tools are generally not used by professional drywall finishers.

Dress Properly

Always wear the proper clothing and safety gear when finishing drywall. Wear goggles and a dust mask to keep dripping compound and dust out of your eyes, nose, and mouth. Adequate clothing can help protect you from cuts and falling objects.

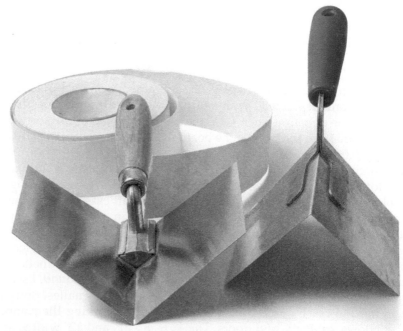

Figure 4 Corner trowels.
Source: sockagphoto/Shutterstock

Some corner systems have a paper-faced corner bead that can be coated with joint compound for direct application to a corner. Corner beads can be screwed on, clinched, or applied with staples.

For attaching corner bead to outside corners, the corner clinching tool may also be used. The clincher is struck with a rubber mallet once the corner bead is positioned, and the tool is set over it. When struck, the corner clinching tool centers the bead and clinches it to the corner by crimping each side into the wallboard at four points. The entire length of corner bead can be attached by moving the clincher up or down the corner and striking it several times (*Figure 5*).

Figure 5 Corner clinching tool with a rubber mallet.

A mud pan (*Figure 6*) is simply a long container that holds joint compound (mud) to be knifed or troweled over the wallboard. The pan is commonly made of steel, aluminum, or plastic. It also may have replaceable steel edges mounted on the long sides for scraping excess compound off the knife or trowel. A typical mud pan looks just like a baker's bread pan.

Sanding is an important part of the finishing process, and tools are available to help you with this task. There are two basic types of sanders: the pole sander (*Figure 7*) and the hand sander (*Figure 8*). Both use the same principle of fastening down a strip of sandpaper by means of clamps and wing nuts.

With the hand sander, you are better able to sand dried finished joints and fastener heads that are within normal reach. With a pole sander, you can do the same job over your head without relying on scaffolds or stilts.

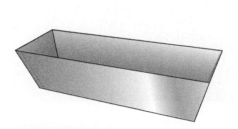

Figure 6 Mud pan.

Figure 7 Pole sander.
Source: valentyn semenov/Alamy Stock Photo

Figure 8 Commercial hand sander.
Source: Vangelis_Vassalakis/Shutterstock

Hand sanders are also known as sanding blocks. A commercial sanding block consists of a wooden or metal base, around which a sheet of sandpaper is wrapped. A second block is pressed against one side of the base and tightened down with a wing nut to hold the sandpaper in place around the base. The sander allows you to apply even pressure over the sandpaper's entire face while saving your fingers from abrasions and other injuries.

Power-driven vacuum sanders, which will be discussed later, are also available.

1.1.3 Mixing Tools

Both hand tools and attachments to power drills can be used to mix joint compound and other liquified materials at the jobsite. For hand mixing, you can use a mud masher, which is a lot like a potato masher. The joint compound (mud) is mashed in a pail until it is mixed to a smooth consistency.

If you need to mix a full bucket of compound, it is faster and easier to use a power drill equipped with a long-stemmed mud mixer (*Figure 9*). There are several types of spinning mud mixers. The mixer end of the device is placed in the joint compound once its shaft has been secured in the power drill chuck. The drill is used like a kitchen hand mixer to stir the joint compound.

1.1.4 Tape Dispensers

A 500' roll of joint tape can be awkward to carry around while trying to apply compound or tape a long joint. Simple tape dispensers and tape holders solve the problem. Many of these lightweight tape-holding devices are designed to hang from belts or shoulder slings, leaving your hands free (*Figure 10*). Some dispensers even crease the tape for application in corners.

A banjo (*Figure 11*) is a large tape dispenser. It is loaded with a full 500' roll of joint tape and a supply of joint compound. The banjo applies the joint compound to the tape as the tape is pulled out. The banjo can be adjusted to change the amount of joint compound applied to the tape. The banjo may get its name from its similarity in shape to the musical instrument of the same name.

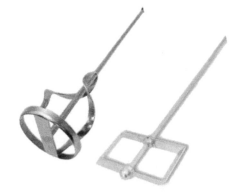

Figure 9 Mud mixer used with a power drill.

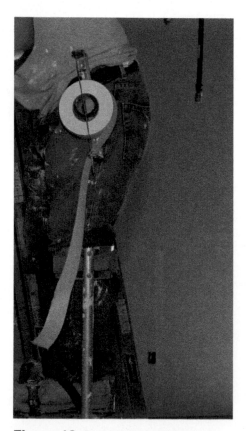

Figure 10 Tape dispenser used with paper or fiberglass mesh tape.
Source: Jim West/Alamy Stock Photo

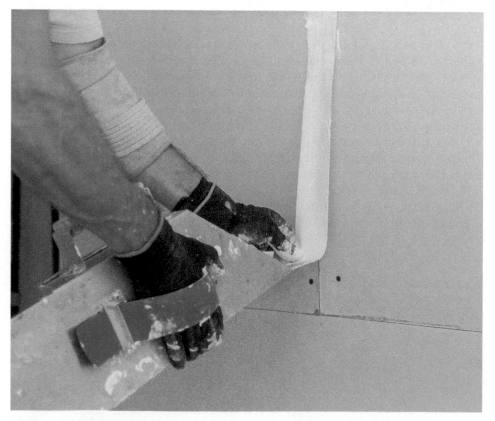

Figure 11 Banjo.
Source: valentyn semenov/Alamy Stock Photo

1.1.5 Automatic Finishing Tools

Tools are available not only to help you apply compound and tape and finish the joints, but also to do many of these operations at the same time—automatically. These tools are operated by hand, but they are referred to as automatic finishing tools or mechanical finishing tools. This is because they use intricate mechanisms to do work that otherwise would have to be done with hand tools and fingers. Although they are operated by hand, the hand operation usually involves simply pushing the tool along a path.

Automatic Taping Tools

One of the most popular mechanical finishing tools is the automatic taping tool (commonly known as a BAZOOKA®), which applies joint compound and tape to joints (*Figure 12*).

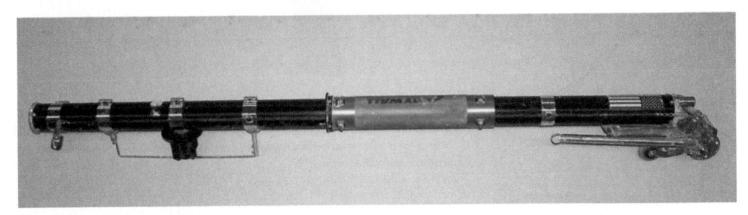

Figure 12 Automatic taping tool.
Source: Kevin Howser

The automatic taper uses gears, rollers, pulleys, a piston, and levers to quickly guide the tape, coat it with a measured layer of joint compound, and dispense the tape along the joint. The automatic taper even cuts the tape at the end of a pass and can crease the tape for application in corners.

Using the BAZOOKA®, ceilings up to 10' high can be finished without using a ladder, scaffolding, or stilts. An extension can be mounted to the taper's base to reach even higher ceilings. For closets and other tight spaces, miniature tapers are also available.

You may need a lot of practice to master the use of an automatic taper, but you will probably find it time well spent. Using automatic tools, an experienced mechanic can tape an entire room faster than using only hand tools.

Automatic tapers are made primarily of aluminum, plastic, and other rust-resistant materials. This makes it easy to keep the unit clean, which is necessary to guarantee proper operation.

Clean the taper by filling the empty joint compound chamber with water. Then force the water out of the unit through the valve that distributes the compound by moving the floating piston up and down inside the compound chamber. Use a high-pressure water hose to spray clean the gears and valve openings at the head of the taper.

Automatic taping tools use a small razor blade to cut the joint tape at the end of a pass. This blade must be changed occasionally, as must the cable that operates the unit's piston. The piston moves up and down inside the taper's tube, forcing out compound. You can change the blade or cable in just a few minutes, following the instructions that come with the taping unit. Always follow the manufacturer's instructions for maintaining the taper to avoid breakdowns and lost time on the job.

Nail Spotters

A much simpler automatic finishing tool is the nail spotter (*Figure 13*). The nail spotter quickly applies compound over nail and screw dimples in fastened wallboard.

Keeping Your Compound Fresh

To prevent the compound from drying out in the taper's head during use, always stand the unit headfirst in a pail of clean water whenever you must stop working for a short time. At the end of the workday, pump all the compound out of the taper before cleaning it out with water.

Figure 13 Nail spotter.

The spotter is simply a metal box on a swiveling pole. The pole's swivel allows you to use the spotter at any angle and to push the device along a wall or ceiling while standing still.

The pole is attached to a hinged plate on top of the spotter. As you push the spotter along, the pressure you exert on the pole is transferred to the plate. As the plate sinks into the metal box, it forces compound out through a small opening in the bottom of the box. The compound fills the dimples, and excess joint compound is automatically scraped off by a blade mounted in the trailing edge of the box.

Nail spotters are commonly made in 2" and 3" widths. The pole lets you reach high ceilings and walls without a ladder or stilts. The mechanism is small and light enough to be used in closets and other cramped spaces.

The nail spotter is much easier to use than the automatic taper. Once it is mastered, an entire room can be spotted in just a few minutes. Because the tool automatically scrapes off excess joint compound and feathers the spot, each dimple must be gone over only once.

The spotter's blade is bowed slightly to leave a small crown of compound over each dimple. Although the blade is very hard, it can be damaged or broken by an exposed nail or screw head. Broken or worn blades must be replaced. The blade is usually mounted to the unit with a clamp and one or two wing nuts. By removing the nuts, you can back off the clamp and slide out the blade. Then simply slide the new blade into place and hold it down by the wing nuts.

The nail spotter should be cleaned thoroughly after every use by flushing with a high-pressure water hose. The units are usually made of rust-resistant aluminum or stainless steel.

Flat Finishers

The flat finisher is also known as a box. You will generally use more than one box because they come in different sizes (*Figure 14*). The idea of a flat finisher is to apply topping and finishing coats to taped drywall joints. As each successive layer is applied, you use a wider box. The flat finisher works on the same principle as the nail spotter and dispenses an even, measured strip of compound along any flat joint. Boxes come in 7", 10", and 12" widths. Each applies topping coats of those dimensions. Use the smallest box for the first topping coat after taping; use the 10" or 12" box for the second topping coat if two topping coats are all that are required; or use the 10" for the second coat and the 12" for a third coat. The 12" box can also be used for applying a fourth or **skim coat**.

Like the nail spotter, the flat finisher is a metal box with a hinged lid. When you push the box along the joint, pressure on the handle forces the hinged lid down, squeezing joint compound out of the box through a small opening at the bottom.

Skim coat: A thin coat of joint or topping compound that is applied over the entire drywall surface. Sometimes required under a high gloss finish.

Figure 14 Flat finishers.

The opening can be adjusted to change the amount of joint compound released. Automatic-feed versions supply compound to the finisher, reducing the amount of effort required.

Once flat joints have been taped either by hand or with an automatic taper, they can be finished using an automatic finisher. This tool applies a smooth, even coat of joint compound over the taped joint, automatically feathering the edges and crowning the center.

The flat finisher operates on the same principle as the nail spotter. The device is pushed along, and pressure on the handle forces compound out of a metal box. As you guide the unit along the joint, a metal blade scrapes off excess compound and leaves the desired crown height. The finisher can be adjusted for different crown heights.

Flat finishers can be used on wall and ceiling joints running in any direction. The tool is not only easy to operate, but also allows you to get the job done in a small fraction of the time it would take to finish joints by hand.

Adjustable handles enable the finisher to reach ceilings up to 12' high. The device is compact enough to use in closets and other cramped spaces. Like most other automatic finishing tools, the flat finishing tool must be kept clean in order to work properly. These applicators must be cleaned thoroughly after each use by squeezing the remaining compound out of the reservoir and flushing the unit with water from a high-pressure hose.

Corner Applicators and Finishers

The corner applicator (*Figure 15*) works in conjunction with a corner finisher, or plow (*Figure 16*). By attaching the plow to the ball/cone end of the corner applicator with a locking retainer clip, the operator can put a finish coat on both sides of an angle at the same time. The plow operates on the same principle as the flat finisher or nail spotter, but its dispensing surface is shaped in a 90-degree angle to fit into interior corners. The applicator is available in 2" or 3" widths and can be used to apply bedding coats of compound prior to taping or to apply finish coats that are feathered and smoothed with other tools.

Once a bed of compound and tape has been applied in interior corners, the tape is smoothed and embedded into the joint compound using a corner roller (*Figure 17*). This device consists of four stainless steel rollers mounted on a swiveling pole.

For best results start from the middle of the angle joint and use light pressure to roll toward both ends. Make a second pass, again from the middle, working toward both ends with firm pressure. This will force excess compound from under the tape. To maintain the corner roller, spray it with a high-pressure hose after use.

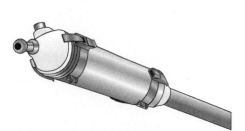

Figure 15 MudRunner™ corner applicator.

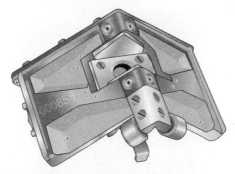

Figure 16 Corner finisher (plow).

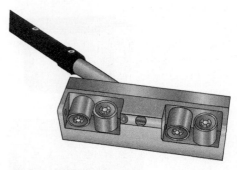

Figure 17 Corner roller.

Automatic Loading Pumps

Automatic taping and finishing tools are filled with compound through automatic loading pumps (*Figure 18*). These pumps look like the old-fashioned hand-cranked water pumps used before the invention of the faucet, and they work on the same principle.

The pump's intake nozzle is placed in a five-gallon pail of compound. To stabilize the device, place your foot on the foot plate on the pump's main body, which sticks out of the bucket and sits on the floor.

As you operate the pump's handle, joint compound is forced out through a J-shaped outlet nozzle. Several attachments can be used to adapt the nozzle to feed various automatic tools. The attachments let you fill the tools quickly and without any over-spill or mess.

Like all automatic tools, the pump must be maintained properly. At the end of the day, remove the device from the pail and force out any compound remaining inside the pump. Flush the pump with water from a high-pressure hose or by placing the intake nozzle in a bucket of fresh water and pumping the water through the compound.

To prevent clogging the pump, make sure the joint compound is properly mixed and free of lumps. Remix the joint compound periodically if it must stand unused for a while.

The pump's intake nozzle features a small screen that stops lumps or debris from being drawn into the pump. Following the manufacturer's directions, remove and clean the screen after each use. You will probably need to replace the screen occasionally.

Priming the Pump

A new or fresh compound pump should be primed by pouring $\frac{1}{2}$ cup of water into the outlet.

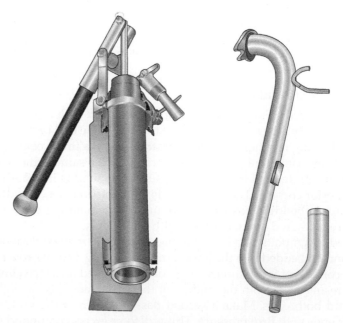

Figure 18 Automatic joint compound loading pump and gooseneck.

Vacuum Sanders

The hand-operated vacuum sander works like a hand-operated pole sander. The vacuum sander's pole, however, is a rigid hose connected to a powerful vacuum cleaner. A piece of screen-backed sandpaper or a tough sanding mesh is stretched across the hose's flat, hollow head. As the sanding head is pushed back and forth across the dried joint compound, the vacuum sander pulls dust and chunks through the sanding mesh and into the hose. The dust is collected in a large filter bag, as in an ordinary household vacuum cleaner.

There are also power-driven versions of these sanders (*Figure 19*). They provide a fast, clean way of finishing drywall jobs. The vacuum hose is connected to a shop vacuum. The sander uses foam-backed sanding pads that are specially designed for drywall work. The machines are safe, effective, and easy to use. Hoses and attachments enable the finisher to sand normal-height ceilings and walls without stilts or ladders.

Figure 19 Power-driven vacuum sander.
Source: Dmitry Kalinovsky/Shutterstock

Dust-Free Sanding

The vacuum sander, or dustless sander, is very useful in environments where dust control is important. Such a device may be required for drywall finishing jobs in hospitals, supermarkets, clean rooms, and food preparation areas. The OSHA silica dust standard for construction (29 *CFR* 1926.1153), requires limited exposure to silica dust.

Source: Pawel_Brzozowski/ Shutterstock

Be sure to follow the manufacturer's directions when using a vacuum sander. Empty or replace the filter bag as often as required and see that hoses are kept clean and sanding materials are in good condition. Although the vacuum sander minimizes free-floating dust, you should still wear appropriate clothes and safety equipment, including goggles and a dust mask.

U.S. Gypsum offers a joint compound that is designed to produce less dust than other compounds. When sanded, it produces larger particles that fall to the floor instead of clouding the air.

1.2.0 **Finishing Materials**

This section covers various types of drywall finishing materials.

1.2.1 Joint Reinforcing Tape

Three kinds of tape may be used in drywall finishing: paper tape, fiberglass mesh tape, and metal edge tape. Each kind may be divided further into those pre-coated with adhesive and those without. The tape most frequently used is paper tape without adhesive.

Standard paper tape is used to cover and reinforce seams, joints, and patch-work. It is a strong paper with feathered edges. There are two types of paper tape available: plain paper and perforated paper.

Paper tape, *Figure 20*, can vary in width, but it is generally about 2" wide. The specific width required by most automatic taping tools is $2\frac{1}{16}$". The tape usually comes in rolls that range in length from 60' to 500'.

The paper tape may also contain fibers, crisscrossed or woven into the material, and it often comes tapered or feathered at the edges. The overall surface may be scuffed or roughened to allow better bonding with the taping compound. Paper tape is considered to be superior to fiberglass tape for many applications because paper tape resists stretching and distortion. It also provides a more consistent bond between the face papers of the gypsum boards on each side of most joints.

Perforated paper tape has larger, more visible holes, which are designed to allow more compound to ooze through them, producing a better bond once dry. These holes may vary in size, depending upon the manufacturer, but they are usually about $\frac{1}{16}$" in diameter.

Fiberglass mesh tape (*Figure 21*) is generally self-sticking fiberglass joint tape made of fabric-woven filaments that do not decay. This type of tape is also available without adhesive. Fiberglass mesh makes a strong tape and may be more durable than paper tape under certain conditions, such as in high-moisture areas and when used with moisture-resistant drywall. Fiberglass mesh tape is much more costly than paper tape, but is good for repair work, veneer taping, and other specialized applications. Its use can increase productivity enough to offset the extra cost of the tape.

Metal edge tape (*Figure 22*) is a type of paper tape with the added feature of two galvanized strips of steel down the center, with a small gap between them to allow for the crease. The metal strips are typically $\frac{1}{2}$" wide. This tape is sometimes referred to as *flexible metal corner reinforcing tape*. However, the tape is still applied and finished just like regular paper tape.

Metal edge tape is best used for corners with other than 90-degree angles. It is also used for corners formed by radius wall and ceiling intersections, arches, drops, splays, and wherever wallboards need to join in unusual configurations. Metal edge tape makes any outside angled corner straight and sharp, with some reinforcing qualities. The edges beyond the metal strips are feathered like those of standard paper tape.

Taping Corners

The center of all paper tape is scored (running the length of the tape) to allow the tape to be easily creased for use at inside corners. The crease side is always the side you press into the corner. In other words, if you fold the tape like a V, the point of the V always goes against the wall. For outside corners, you generally use corner bead.

Source: darkwisper/123RF

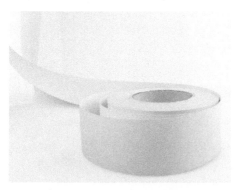

Figure 20 Joint reinforcing tape.
Source: Kellis/Shutterstock

Figure 21 Fiberglass mesh tape.
Source: Anton Starikov/Alamy Stock Photo

Figure 22 Metal edge tape.

Using Fiberglass Tape

Fiberglass mesh tape pre-coated with adhesive is designed to be installed without first applying a bedding coat of taping compound. In other words, adhesive-backed tape is applied directly to the joint and pressed into place with a taping knife or trowel. After the tape is stuck into place, it is covered with layers of compound. Fiberglass mesh tape without adhesive backing is sometimes installed using staples.

Some manufacturers recommend that you do not use fiberglass mesh tape with the usual ready-mixed or powder joint compounds. Instead, special powder compounds, such as quick setting compounds, have been developed that work much better with fiberglass tape. Quick setting compounds are discussed in more detail in this module.

Corner bead is not very flexible, so something else is needed for odd corners made by curved or angled wallboard intersections. The principal idea of metal edge tape is to provide, as much as possible, the same strength and finishing quality that corner bead provides.

Metal edge tape is generally $2\frac{1}{8}$" wide and comes in 100' rolls. The gap between the metal strips, allowing for the paper crease, is usually only $\frac{1}{16}$". The tape is designed so that the metal side is applied to the wall.

1.2.2 Finishing Compounds

Joint compound is more commonly known as mud. It comes in both wet and powder forms. Wet compound has been premixed by the manufacturer, while powder compound is mixed on the jobsite by adding the proper amount of cold, clean water. Each of these products is designed to accomplish a specific result, depending on the job. Their ingredients are often different, so you should never mix wet compound with dry compound.

The two basic types of joint compound are setting-type and drying-type. Drying-type compounds (*Figure 23*) have a consistency that makes them easy to sand. However, they typically take 24 hours or more to fully dry so they can be sanded and finished. Drying-type compound is commonly used in high-end work. Drying-type compounds are available in powder and pre-mix forms.

Setting-type compounds (*Figure 24*), also known as quick setting compounds, contain chemicals that cause them to harden quickly. Although they are convenient because of their quick-dry qualities, these compounds do not sand easily. They must be smoothed before they completely harden. Quick setting compounds are often used for embedding tape, followed by finish coats of drying-type compound. Because these compounds begin to harden as soon as they are exposed to air, they are only available in powder form.

Several types of joint compound (mud) are used in finishing drywall. One popular approach is the two-step system in which the tape is embedded in the joint using a taping compound to obtain a strong bond. The joint is then finished with a topping compound, which is much easier to sand. An **all-purpose compound** combines the characteristics of the two-step system into a single compound.

Taping compound is designed for its bonding qualities and strength in bedding and reinforcing taped joints. It is also used as a first coat on metal corner bead or trim, nail or screw heads, and other fasteners. Taping compound is also used as prefill and fill coats, and for repairing surface drywall cracks and cracks in plaster. Taping compound is generally the most likely to shrink. It is also the strongest bonding and most difficult to sand.

Topping compound is used for the second and third coats of the finishing process. This type of compound is softer and easier to sand. It also produces less shrinkage. You must never use topping compound for taping because topping compound is not designed to embed and bond a taped joint. A joint bedded with topping compound will crack with the first slight movement or settling of the wall. Topping compound is designed to be molded and sanded flat on top of a joint that has already been fastened together.

All-purpose compound combines the features of both taping and topping compounds, but in so doing, it gives up some of the qualities of each.

Figure 23 All-purpose drying-type compound.
Source: Steve Cukrov/Shutterstock

All-purpose compound: Combines the features of taping and topping compounds. It does not bond as well as taping compound, but finishes better.

Figure 24 Setting-type compound.

Figure 25 Lightweight compound.

Lightweight compound: An all-purpose compound having less weight than standard compounds.

For example, it loses some of the bonding qualities of taping compound and some of the soft and smooth drying capabilities of topping compound.

However, all-purpose compound is excellent for use in textured finish applications. This type of compound is often used to finish walls with various interesting effects. In that respect, you are almost practicing the art of plastering.

Lightweight compound (*Figure 25*) has the advantages of an all-purpose compound and is 25 to 35 percent lighter. It also has less shrinkage and sands as easily as topping compound. Lightweight compound can be used to laminate gypsum panels, coat interior concrete ceilings and above-grade columns, and patch cracks in plaster. It can also be used for texturing.

Powder Compounds

Compounds packaged in dry powder form store better than other forms of compound. Dry powder can be stored at any temperature and in any storage area or warehouse that is kept dry and free from moisture. Because warehouses are rarely heated, powder compounds are best for winter storage. However, it is recommended that powder compounds be moved to a warm mixing room a full day before they are mixed.

All powder compounds must be mixed with clean water in exactly the proportions specified on packaging instructions. Generally, these proportions depend upon the amount of compound you need for the job. The amounts needed are also determined by the square footage of the joint areas you intend to cover. Once mixed, the compound may be kept in tightly covered containers in storage for many days, if the storage area is kept at room temperature. If the powder has been properly mixed to start with, it will not clump up or harden in stored closed containers, but you should always stir or mash it again before use.

Both powder and premixed forms of compound shrink by drying out. As water evaporates from the compound, the compound shrinks to fill in space vacated by the water. This process is different from the chemical hardening process used by special quick setting compounds, which is more like glue drying.

Taping, topping, and all-purpose compounds are generally available in dry powder form. They may be packaged in bags or cartons. The bags generally contain 25 pounds of powder. Cartons may be measured in gallons or pounds and usually contain more powder than bags.

Current building codes and standards generally forbid any use of asbestos in construction materials. Always make sure that the powder you are going to use is right for the job and complies with local regulations.

When Is Five Gallons Not Five Gallons?

The traditional five-gallon pail is no longer five gallons. Currently, these full plastic pails contain only 4½ gallons of premixed compound.

Working with Powder Compounds

Observe the following practical rules regarding powders:

- Always use clean, cold water for mixing.
- Never vary from the correct proportions of water-to-powder specified for the product you are mixing.
- Always label and date the container you use for mixing. Some drywall mechanics write a description of the contents and the date on a piece of masking tape and attach it to the lid. This is generally better than using ink or permanent marker directly on the lid because you can pull the tape off and use the lid again later.
- Never confuse the different mixes. Keep taping compounds separate from topping compounds and so forth. If everything is labeled, you will not have this problem. (Hint: Mix powder taping compound only in empty premix taping compound pails, and mix powder topping compound only in empty premix topping compound pails.)
- Be sure to use the oldest dated compound first before mixing a new batch.
- Always avoid mixing one brand of powder with a different brand of powder. Even if they are supposed to be the same kind of powder, different manufacturers make their products differently. They are usually not compatible with similar products made by other companies.
- Follow the dry mix safety instructions given in this module.

Premix Compounds

In general, premix (ready-mix) compounds are formulated, mixed, and packaged by their manufacturers. They are vinyl-based and require little or no mixing. This feature reduces the need for readily available water on the jobsite. Premix compounds also reduce the waiting or soak times after application, and they offer good crack resistance after drying.

These products will freeze, however, so you need to take precautions in cold weather. If vinyl premix does freeze, thaw it only at room temperature, and do not apply any additional heat. Always use pre-mix products in their packaged state of consistency as much as possible to minimize shrinkage. Follow the specific instructions on the label.

Premix compounds are available in both plastic pails and sealed cartons. They are also quite heavy because, as the name implies, the water is already mixed in. The full pails weigh over 60 pounds; full cartons might be 50 or 60 pounds.

The advantage of premix compounds is that they can be used at jobsites where there is no supply of fresh water, which must be available in order to use powders. Of course, powders can be mixed where there is fresh water and then transported in closed pails to sites without water. Generally, however, most contractors simply use the premix in these situations.

The disadvantage of premix is that it must be stored where it will not freeze during the winter. Finishing compounds must be used only in room-temperature environments. In colder climates in winter, you cannot do drywall finishing until the jobsite interiors are closed up and heat is installed and working.

Even though the premix is already mixed at the factory and ready to use on the job, you will want to mix or mash it again before you use it. Hand mashing is done by forcing the masher down repeatedly in the center and around the sides of the pail. Use smooth, complete downward strokes. On the upward stroke, scrape the masher along the sides of the pail to loosen any compound sticking to the edge.

Only two or three minutes of mixing is usually enough to ensure a smooth, even consistency all through the compound.

You will need to dilute the premix compound slightly for use with automatic finishing tools. Generally, add $\frac{1}{2}$ cup of clean, cold water to $4\frac{1}{2}$ gallons of premix compound. For the automatic taper, mix in two cups of clean, cold water per $4\frac{1}{2}$ gallons of compound. Note that some manufacturers have compounds specifically designed for use with their line of taping and finishing tools.

The full range of compounds are available in premix form: taping, topping, all-purpose, and specialty compounds. Specialty compounds are all-purpose compounds that offer enhanced bonding, shrinking, and sanding capabilities. They often eliminate the need for a third topping coat, and they are good for laminated applications. Specialty compounds are also useful for texturing applications.

Setting-Type Compounds

Quick setting compound, commonly called 20-minute mud, 30-minute mud, or hot mud, is a compound that sets up very quickly in comparison to other compounds because it hardens chemically instead of by water evaporation. Shrinkage is reduced considerably, so quick setting compound makes an excellent filler for metal trim, repairs, and around pipes.

Quick setting compound is available only in powder form; it needs to be mixed with water on the job. It is essential that the water be clean and cold. It is also very important to mix only as much as you can apply in the allotted setup time.

Quick setting compound is good for small jobs, pre-fill, corner beads, and finishing bathrooms and other high-moisture areas. It is especially good in humid weather. Because it sets so quickly, you need to wipe off excess compound immediately. Sanding the dried compound is also difficult, so take care to apply and wipe it as smooth as possible before it dries. Accelerators are available to make quick setting compounds set even faster for special needs.

Quick setting compound is packaged and sold by its setup time, which generally ranges from 20 minutes to 360 minutes (6 hours). Note that the compound

Quick Setting Compounds

A disadvantage of quick setting compound is that it can be difficult to sand. That is why it is generally used as a prefill or tape-and-bed coat, while regular topping compound is used for second and third finishing coats. Quick setting compound should not be used for the finish coat because it might bleed through the decorated surface. Sandable quick setting compounds are available.

Figure 26 Quick setting compound.

> **NOTE**
>
> Don't over-mix setting compounds. Over-mixing will cause air bubbles in the finish.

shown in *Figure 26* sets up in 45 minutes. Because it works chemically, once it is hardened it will not shrink, even though it is not dry. This allows a strong bond to form and remain in high-humidity environments.

Quick setting compounds are also very good for laminating applications, especially for laminating wallboard layers together. These compounds can be used for coating concrete walls and ceilings (above ground), for filling in cracks and holes, for skim coating, and even for surface texturing. They are also preferred for finishing exterior ceiling boards and for presetting joints of veneer finish systems.

Dry-Mix Safety

Whenever you mix dry compounds, be aware of the dust level and try to keep it to a minimum. You must also use the proper respirator when mixing dry powder compounds. In addition, observe the following guidelines:

- Make sure all mixing containers are clean and free of residue.
- Use only clean, drinkable water for mixing.
- Make sure all tools and mixing blades are clean.
- Mix these compounds only according to the directions on their labels.
- If you use a power mixer or mixers operated by a power drill, use a slow speed such as 300 rpm to 500 rpm.
- Do not try to mix different types of compounds together. Their chemical makeup often differs from manufacturer to manufacturer. Even with the same manufacturer, the different types of compound are not compatible with each other.

Rules for Quick Setting Compounds

Observe the following rules when using quick setting compounds:

- It is best to mix with power mixers.
- Mix only enough to be used in the time stated in the product instructions.
- Clean mixing equipment and tools immediately after using them because hardening will occur even when they are submerged in water.
- Temper the compound.
- Never add water to the compound after tempering.
- Never try to use this compound in an automatic finishing tool, such as an automatic taper.

Estimating Joint Treatments

The approximate quantities of materials needed for 1,000 square feet of wallboard are as follows:

- *Joint tape* — 370'
- *Joint compound* — 83 pounds of conventional powder or 138 pounds of conventional ready-mix

These figures can be used to calculate the requirements for other square footages by reducing the square footage to a decimal percentage of 1,000 and multiplying the area required by the preceding quantities. For example, 2,000 square feet would require twice the amount listed above and 1,400 square feet would require 1.4 times the amount listed above.

1.2.3 Sanding Materials

Sanding operations may be done with sponge sanders, hand sanders, pole sanders, or power-driven vacuum sanders, which were described earlier. The sanders require sheets or strips of sandpaper to be fitted to them. Sanding is an important part of the finishing process, and tools are available to help you with this task. With the hand sander, you are better able to sand dried finished joints and nail spots that are within normal reach. With the pole or power sander, you can do the same job over your head without having to bother with scaffolds or stilts.

Sandpaper is rated by coarseness, called *grit*, which varies by degree of fineness. The lower the grit number, the coarser the paper. Coarse paper is used for first sanding jobs where the surface is very rough and you are trying to smooth it out fairly quickly. Fine paper is used for finish sanding where you are making the surface as smooth as possible, usually in preparation for painting.

Sanding is typically done with 180-grit sandpaper. Sandpaper is also available in grit numbers from 80 to 150 or more. It is packaged in sheets for hand sanders and in discs for power sanders. The precut sheets are designed to fit specialized tools such as angle sanders and wall sanders (sanding poles). Precut sheets are usually sold in packages of 100.

Other sanding materials include sanding cloth, abrasive mesh cloth, and open mesh cloth, which are used chiefly for cutting joint compound. These materials may also be packaged in rolls or as individual sheets. Sanding sponges have become popular for sanding work. Sanding sponges have abrasive material bonded to one or both sides of a sponge. Some of them are beveled to make it easier to sand inside corners.

Mesh cloth, commonly called sanding screen, is preferred by many finishers for use on pole sanders. For first sanding of a joint or seam, this cloth prevents raising the nap (tiny hair-like fibers) on the face paper of the drywall. Do not raise this nap at all, if possible, because a smooth nap makes for better painting. An advantage of mesh cloth is that it does not load.

Film-backed drywall abrasives are also effective. A major advantage is that they do not raise the face of recycled paper.

1.2.4 Textures

A texture is any wall or ceiling coating that serves as its own finish. It may also serve to hide taping and other finishing so that the surface need not be sanded smooth or otherwise prepared for painting, wallpapering, or other treatments.

Decorative textures are very popular in both residential and commercial construction. Interesting patterns, simulated acoustical effects, and light or heavy finishes may be applied with rollers, brushes, stencils, sprayers, and other tools and equipment. The following are some of the more common pieces of equipment used in texture finishing, as shown in *Figure 27*.

1. *Glitter gun* — Used to embed glitter in wet texture ceilings. The hand-crank model shown is most economical, but not as efficient as an air-powered type.
2. *Drywall paddle mixer* — Used with an electric drill at less than 400 rpm to mix joint compound. It is designed to reduce the entrapment of air bubbles in the mixture.
3. *Stucco brush* — Used to create a variety of textures from stipple to swirl. Other variations can be achieved with thicker application and deeper texturing.
4. *Texture brush* — Available in many sizes and styles; tandem-mounted brushes cover a large area to speed a texturing job.
5. *Wipe-down blade* — Has a hardened steel blade and long handle to speed cleaning of walls and floors after application of texture materials. The blade has rounded corners to avoid gouging.
6. *Long-handled roller* — A standard paint roller adapted to the particular type of finish required. Available roller sleeves include short nap, long nap, looped, foam stipple, and carpet types in professional widths.
7. *Texture roller pan* — Used with rollers. Some models can hold up to 25 pounds of mixed texture.
8. *Flat blade knife* — Used to apply texture material and for troweled finishes.
9. *Circular patterned sponge* — Used to achieve patterned swirl finishes.
10. *Sea sponge* — Used to achieve free-form texture finishes.

Wet Sanding

If only minimal sanding is required, a wet sponge can be used. This method produces no dust and will not scuff the paper. Use a polyethylene sponge, which looks something like carpet padding. Wet the sponge with clean water that is cool to lukewarm. Wring out the sponge enough to prevent dripping, then rub the joints to remove high spots using as few strokes as possible. Clean the sponge frequently during use.

Texture Materials

Texture materials are similar to joint compounds in that they are made in both powder and premixed forms. In fact, some joint compounds can be easily used for texturing.

11. *Whisk broom* — Similar to a stucco brush but stiffer. It can be used to produce a bolder brush pattern.

12. *Short-handled roller* — Same as the long-handled roller with the same sleeves. A looped texture roller sleeve is shown.

13. *Texture sprayer* — Used to spray textured finishing materials on drywall.

14. *Texture paddle* — Similar to the drywall paddle mixer, but designed for mixing texture materials at 300 rpm to 600 rpm.

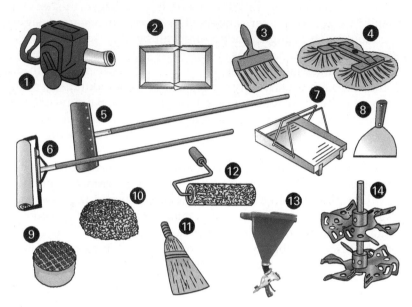

Figure 27 Typical texture finishing equipment.

Spray Texturing Machines

Spray texturing machines are available for large jobs. They range from a hopper with a pneumatic spray nozzle to a self-contained machine like this one with its own built-in compressor.

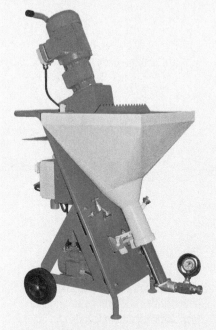

Source: stefan11/Shutterstock

Texture materials may be manufactured in both powder and ready-mixed forms. The four types are described as follows:

- *Powder textures* — These textures may be either aggregated or unaggregated, which means there may or may not be other particles mixed into the powder. In aggregated products, particles of such substances as perlite, vermiculite, and polystyrene are used to make textured effects on primed surfaces, particularly ceilings. Aggregated powder products are designed for spray application. They also have a good solution time, only minimum-to-moderate fallout, and good bonding power and crack resistance. When properly sprayed on, these textures hide substrate imperfections very well. Unaggregated powder products may either be sprayed or hand applied to primed walls or ceilings. Several of these products are limited to hand applications only, so that you can produce crow's foot, stipple, or other pattern texture effects. Crow's foot is a design produced by a roller, which makes a kind of random bird-track pattern across the textured surface.

- *Powder joint compound textures* — Generally, powder joint compound textures are the same as the topping or all-purpose powder compounds used for normal joint finishing. For texturing with these products, you use a brush, roller, or trowel to produce light and medium hand-formed textures on walls or ceilings. Typically, you use swirling motions to make random patterns. Powder joint compounds are easily mixed in the usual way. They are smooth working and easy to texture by hand. They also hide surface imperfections very well, produce good bonding, and resist cracking. The color is white after drying, which can be left as is or painted another color, depending upon the finish specifications.

- *Premixed textures* — These include thick, heavy-bodied, vinyl-based materials, which can produce smooth, very deep textures. They generally dry to hard white finishes that are often left unpainted, especially on ceilings.

These textures may be sprayed, troweled, or applied with a roller or brush. They go well over concrete and can fill voids and cracks and cover surface blemishes. Premixed texture offers good resistance against cracking on walls and ceilings. These textures are factory-mixed to a smooth consistency. They are easy and fast to apply and generally produce favorable results.

- *Premixed joint compound textures* — These textures consist of topping or all-purpose compounds, which can produce light to medium textures on ceilings and walls. These compounds are well suited for small jobs that need only brushes, rollers, or trowels. However, they can also be spray applied. You can use these materials to produce a great variety of patterns and designs. These textures dry white, offer good crack resistance, and can be painted when dry. They are factory mixed to be smooth and free from lumps. These compounds go on quickly and easily for low-cost, yet good quality results.

1.0.0 Section Review

1. The _____ applies joint compound to the tape as the tape is dispensed onto the wallboard.
   a. corner trowel
   b. banjo
   c. finishing knife
   d. tape dispenser

2. _____ is generally self-sticking tape made of fabric-woven filaments that do not decay.
   a. Paper tape
   b. Perforated paper tape
   c. Fiberglass mesh tape
   d. Metal edge tape

2.0.0 Drywall Finishing Procedures and Standards

Objective

Explain drywall finishing procedures and describe finish levels.

a. Explain how site conditions affect a drywall project and describe drywall inspection.
b. Describe the steps of the drywall finishing process.
c. Describe automatic taping and finishing procedures.
d. Describe hand finishing procedures.

Performance Task

3. Distinguish a finish level by observation and identify the steps needed to take it to the next level of finish.

Drywall finishing involves taping, topping (also known as *buttering*), and sanding the wallboard seams and joints, whether on walls or ceilings, so these surfaces can be made ready for final decorating. Some jobs use texturing and have a reduced need for taping and sanding. Other jobs require a large amount of butt and seam taping, three or more topping or skim coats, and a lot of sanding to produce the expected results.

Drywall requires different levels of finish depending on its location, lighting conditions, and decorative treatment. A hidden surface, such as an attic area, requires far less finishing than one in full view, such as a living room wall. In addition, even minor flaws are apt to be more evident when the surface is exposed to strong lighting conditions or decorated with certain finishes, such as gloss or semi-gloss paints or thin wall coverings. Generally, the more visible a surface, the more likely its lighting or decoration are to show surface defects, and the more finishing work it requires.

These factors are addressed in a document titled *Recommended Specification on Levels of Gypsum Board Finish*, which was jointly developed by the Painting and Decorating Contractors of America, the Association of the Wall and Ceiling Industries—International, the Gypsum Association, and the Ceilings & Interior Systems Construction Association.

This specification is designed to serve as a standard reference for architects, specification writers, contractors, building owners, and others. It provides them with a specific description of the final appearance of gypsum walls and ceilings finished to different levels before the application of a decorative coating of paint, texture material, or wallcovering.

The specification describes the following six levels of finish and typical applications for each of them:

- *Level 0* — No taping, finishing, or accessories required. This level might be used for temporary construction or where final decoration is undetermined.

- *Level 1* — All joints and interior angles shall have tape embedded in joint compound (also referred to as *mud* or *taping compound*). Surface shall be free of excess joint compound. Tool marks and **ridges** are acceptable. A Level 1 finish might be specified for attics, areas above ceilings, service corridors, and other areas not generally seen by the public. It provides some degree of smoke and sound control. In some areas, it is called *fire-taping*.

- *Level 2* — One separate coat of joint compound shall be applied over all joints, angles, fastener heads, and accessories. The surface shall be free of excess joint compound. Tool marks and ridges are acceptable. All joints and interior angles shall have tape embedded in joint compound. Joint compound applied over the body of the tape at the time of tape embedment shall be considered a separate coat of joint compound and shall satisfy the conditions of this level. A Level 2 finish might be recommended for garages, warehouses, and other areas where surface appearance is not critical. It is specified where water-resistant gypsum backing board *(ASTM C630)* is used as a substrate for tile.

- *Level 3* — All joints and interior angles shall have tape embedded in joint compound and one separate coat of joint compound applied over all joints and interior angles. Fastener heads and accessories shall be covered with two separate coats of joint compound. All joint compound surfaces shall be smooth and free of tool marks and ridges. A Level 3 finish might be specified for surfaces to be finished with a medium or heavy texture before painting or with heavy-grade wallcovering. It is not recommended for light or medium weight wallcoverings or for smooth painted surfaces.

- *Level 4* — All joints and interior angles shall have tape embedded in joint compound and two separate coats of joint compound applied over all flat joints and one separate coat of joint compound applied over interior angles. Three separate coats of joint compound shall be applied over all fastener heads and accessories. All joint compound shall be smooth and free of tool marks and ridges. Light textures or wallcoverings require this level of finish. The specification notes that in critical lighting areas, flat paint over light textures reduces shadowing of finished joints through the surface decoration, but that gloss, semigloss, and enamel paints are not recommended for this level of finish. It also notes that the type of wallcovering applied over this level should be chosen carefully to properly conceal joints and fasteners.

- *Level 5* — All joints and interior angles shall have tape embedded in joint compound, two separate coats of joint compound applied over all flat joints, and one separate coat of joint compound applied over interior angles. Three separate

Ridges: Slight protrusions in the center of a finished drywall joint that are usually caused by insufficient drying time. Also known as *beads*.

coats of joint compound shall be applied over all fastener heads and accessories. A thin skim coat of joint compound, or a material manufactured especially for this purpose, shall be applied to the entire surface. The surface shall be smooth and free of tool marks and ridges. Level 5 is recommended for gloss, semigloss, enamel, or nontextured flat paints or severe lighting conditions. This is the highest level of finish and provides the best protection against joints or fasteners being visible through the decorative coating.

Finishing Steps

Paper tape and compound are applied to a drywall joint in the following four basic steps:

Step 1   A coat of taping compound is applied to the joint equally on both sides. This tape is one-sided. The exterior fold of the crease should be applied toward the gypsum board.

Step 2   Paper tape is pressed into the joint and is smoothed with a 4" knife. (When automatic taping equipment is used, the compound and tape are applied in one step.) Remove excess compound and skim the tape lightly to conceal wrinkles.

Step 3   When the first coat of compound is dry, a coat of topping or all-purpose compound is applied with a 6" taping knife and feathered out 7" to 10" on either side of the joint with a finishing knife.

Step 4   When the second coat is dry, a third coat is applied and feathered out at least two inches wider than the second coat.

The specification, which is available from any of the associations that developed it, notes that for Levels 3, 4, and 5, it is recommended that the prepared surface be coated with a drywall primer prior to the application of finish paint. It also notes that the effects of severe lighting on a surface can be minimized by skim coating the drywall, by decorating it with medium to heavy textures, or by using window coverings to soften shadows.

2.1.0 Site Conditions

Jobsite conditions such as temperature and humidity affect the performance of most finishing materials. During winter conditions, drywall finishing should not be attempted unless the building has heat in a somewhat controllable range between 50°F and 80°F. Furthermore, all materials must be protected from the weather at all times. If the humidity is excessive, ventilation must be provided. Windows should be kept open to provide air circulation. In enclosed areas without natural ventilation, fans should be used to move the air. When drying is slow, additional drying time should be allowed between applications of joint compound. During hot, dry weather, drafts should be avoided so the joint compound will not dry too rapidly.

2.1.1 Drywall Inspection

The professional always inspects installed drywall before finishing it. This determines if the drywall is ready for joint treatment. Improperly installed drywall is difficult to finish.

Examine the hung drywall. Nail and screw heads should all be dimpled. This means that each fastener head should have made a slight depression when it was driven into the wallboard (*Figure 28*). No part of the fastener head should be above the rest of the board's surface. You can check this easily by running your bare hand over the rows of fasteners.

Determine if the fasteners are holding the wallboard panels tightly against the framing members. You can detect loose fasteners by placing a finger over the fastener head. Press the adjoining drywall area in toward the framing with

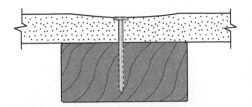

Figure 28 Uniform depression.

your other hand. When loose, movement will be felt through the fastener head. If you discover this, install an extra fastener above and below the original fastener to make the drywall more secure.

Check the butted joints and outside corners. Remove loose paper and broken board from the drywall edges. Cut the paper back to where it still adheres; do not pull it.

Examine the board fields and the butted joints. Check for torn areas and large gaps. Mark all areas with a lead pencil only; ink or crayon will bleed through. These damaged areas must be repaired.

Examine the inside and outside corners. How well do the wallboard panels butt together? Make sure the panels are properly aligned. When one board sticks out farther than the other, the joint is difficult to tape and finish smooth. Another way of saying this, especially for butt joints, is that there might be a high side and a low side. A high side is produced by a butted panel that sticks out too far from the framing. A low side is then produced in the other abutting panel, which does not stick out as far. This is shown in *Figure 29*. This condition might be caused by a twisted stud. Part of the panel rests against a part of the stud that is not even with the part of the stud to which the other board is attached.

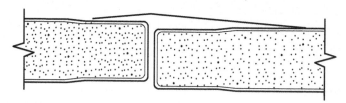

Figure 29 Butt joint misalignment (low side on right).

Since at this stage it is too late to correct the framing, the best you will be able to do is hide the offset. You can hide it by applying a little more compound and finishing the joint a little wider than usual. In fact, any butt joint at all is finished wider than a **tapered joint** because there is no tapered depression to hold the compound. The best you can do with any bad butt joint is to camouflage it.

All these inspections help you determine which procedures to use first in any given situation. If a major repair is required, inform your supervisor. If drywall needs serious correction, such as reframing or rehanging, it is much better to get it done before you begin taping and trying to hide everything with compound. However, if the repair is not that serious, you should be able to fix it.

Tapered joint: A joint where tapered edges of drywall meet.

2.2.0 **Overview of the Finishing Process**

As a general statement, the ideal drywall finishing process requires five steps. Depending on the drying time, it may take five different trips to the jobsite:

Step 1 Complete repairs, cutouts, and prefill. Apply bedding tape at all joints and seams. Complete corner bead and trim installations. Top outside corners. Use an 8" knife on headers and spot fastener heads.

Step 2 Apply first topping coat over taped joints and seams. Use a 10" box. Make double-wide topping coats at all butt joints. Top all corners and angles.

Step 3 Sand lightly, and scuff angles and flats. Apply a second (flash) topping coat to fill in all pits, gaps, depressions, or shrinkage. Apply a straddle coat on the butt joints. Apply another coat on the fastener heads.

Step 4 Perform light sanding and scuffing. Apply a third (skim) topping coat. Use an 18" knife on flat seams and butts. Use a 3" plow for angle topping. Apply a flash coat on headers, seams, and outside corners.

Step 5 Complete pole and hand sanding.

Maintaining the Proper Temperature

If drywall finishing is done during cold weather, the building must be heated to 55°F minimum, and the heat must be maintained during the entire finishing process and until the finish is dry. Avoid sudden changes in temperature, which can cause cracking due to thermal expansion.

Finishing compounds lose strength if they are subjected to freeze-thaw cycles. If a compound has been frozen, it should be discarded.

Source: Kitch Bain/Shutterstock

Drying Time

Atmospheric conditions and other factors always play a part in how fast the taping and topping coats dry. Another factor might even be the wallboard face paper itself. New and recycled paper might well have different drying rates, and different compound materials will vary as to how fast they dry. Therefore, drying times might be longer than just overnight.

2.3.0 | Automatic Taping and Finishing Procedures

Taping and finishing is a multi-step process that varies between three and five different stages of finishing of each board joint or seam. Joints are generally considered to be any place where two edges of wallboard come together. Seams are places where tapered board edges meet each other. Butt joints are places where square (non-tapered) board edges meet each other. Flat joints are places where two beveled edges meet.

Generally speaking, joint taping includes pre-filling the joint, taping and bedding, topping and skim coats, and sanding. After each step, the compound is allowed to dry, usually overnight.

Joint compound and tape shrink as they dry. This shrinkage results in slight depressions that need to be filled out again by applying topping, skim, or finishing coats of compound. Using an automatic taping tool system greatly speeds the processes at each joint or seam. The whole idea is to make every joint or seam as flat as the rest of the wall or ceiling. The joints and seams should be undetectable once the decorating is complete.

Large jobs may require drywall finishing equipment. These specialized tools, which were discussed previously in this module, enable drywall finishers to tape and finish drywall uniformly and efficiently. Basic tool components of such a system are the loading pump, nail spotter, automatic taper, corner roller, flat finisher, corner finisher, and corner applicator. The finishing process using automatic tools is outlined as follows:

Step 1 Apply the tape using the automatic taper.

Step 2 Press the tape into corners using the corner roller.

Step 3 Wipe down excess compound and embed the tape using a broad knife or taping knife.

Step 4 After the bedding coat has dried, apply a topping coat using a flat finisher for seams and butts. Use a corner applicator/finisher for angles.

Step 5 After the first topping coat has dried, apply a second topping coat using a wider flat finisher for seams and butts, and a corner applicator and finisher for angles.

Step 6 After the second topping coat has dried, another skim coat may be applied using taping knives.

Step 7 After all coats have dried, sand the areas treated with compound. Wipe them down with a damp sponge after sanding. This helps the paper fibers to lie down.

2.3.1 Pump Loading Procedures

The loading pump (*Figure 30*) has nozzles of different sizes. Nozzle selection depends upon the equipment and material being used. The pump has a replaceable screen at the loading pump intake. This screen prevents large particles from passing through the pump. The pail holder is designed for a standard five-gallon pail. A gooseneck attachment mounts on the pump to fill the automatic taper. The loading pump without the gooseneck attachment fills the nail spotter, flat finisher, and corner applicator.

Thinning the Compound

Experience has shown that the best results come from adding $\frac{1}{2}$ cup of clean, cold water to $4\frac{1}{2}$ gallons of compound for all automatic tools except the automatic taper. For the automatic taper, mix in two full cups of clean, cold water per $4\frac{1}{2}$ gallons of compound. Use the compound full strength for all hand tool applications, nail spotting, prefill, and skim coating.

Figure 30 A drywall taping tool being loaded with an automatic loading pump.
Source: Kevin Howser

The pump is simple and rugged. It is designed to fill automatic taping and finishing tools with compound. It requires very little training for use and fills the application tools quickly and surely without pumping air.

Be sure to mix all compound thoroughly before using and especially before pumping into any automatic tool. Be sure to mix out any lumps, especially when using powder compounds. Use a power drill with mixer attachment, if available, or a mud masher. With a mud masher, you should plunge it down into the center of the pail and bring it up, scraping against the sides. Rotate around the pail as you mix. This keeps lumpy residues from forming on the sides of the pail. Also, be sure to remix any compound that has been left standing for a period of time.

Keep the pump screen clean and free of lumps or dried compound. It may need to be replaced quite often, along with the O-ring on the gooseneck.

A final recommendation is to have two loading pumps at each jobsite, one with the gooseneck attachment, and the other with a fill adapter. This will speed up the operations of filling the different types of automatic tools.

2.3.2 Pumping

Set the pump into a full standard-size premix compound pail, step on the pump's footplate outside the pail, and pump the compound using the pump handle. Before you pump any compound into an automatic tool, however, you may need to add water to improve the consistency of the pre-mixed compound. Taping compound should have a thinner consistency for use with the automatic taper, and a thicker consistency for topping applications. Always follow the recommended mixing instructions on the labels of the products you are using.

Also, before pumping any compound into an automatic tool, be sure to pump the handle a few times to clear out any air. Without the gooseneck attachment, you will simply be pumping compound back into its own pail. With the gooseneck attached, you should use another container to catch these first few pumps. Never attach a taper to the gooseneck until you are sure all the air is out. Then attach the taper upside down to the fitting and pump it full.

2.3.3 Automatic Nail Spotter Procedures

You may use a nail spotter to fill countersunk nail and screw heads with joint compound. A complete row of fastener heads can be filled in one pass with this tool, which is normally available in 2" and 3" widths. The tool allows you to fill

Neatness Counts

Always set up your mixing pails and other equipment on a large scrap sheet of wallboard. This helps keep splashes off the floor and will greatly simplify your cleanup efforts. You should also keep at least one full bucket of water handy in this area to soak automatic tool heads when not in use.

rows of fastener head depressions with compound, on both walls and ceilings, while working from the floor.

Use the loading pump to fill the nail spotter with compound. Set the pump into the pail, making sure the compound is well mixed and free of lumps. The pump needs only to be pumped full of compound; no priming is needed. Use the adapter spout (not the gooseneck) to fill the nail spotter at its opening.

Each row of depressions can be filled in one pass. Be sure to make positive contact with the wallboard surface at the beginning of each row. Draw the tool smoothly along the entire row, applying some pressure to force the compound out onto the surface. The blade skims off excess compound and leaves a slight crown over each depression as the tool floats along on the rocking skid.

After you pass over the last depression, gradually break contact with the surface by using a sweeping motion. This procedure will fill the depressions without leaving excess compound that needs to be removed by hand.

The nail spotter is a simple tool to learn to use. Generally, you will have the procedure down by the end of your first day of using it.

2.3.4 Automatic Taping Procedures

Fill the automatic taper with joint compound using the gooseneck adapter with the loading pump. Close the gate valve and turn the tool upside down to fit its opening over the gooseneck opening. Put one or two fingers of your free hand into the end of the taper. Pump in the compound and stop pumping when your fingers feel the piston. This is to avoid overfilling. If you do happen to overfill, relieve pressure by depressing the filler valve stem with a nail.

WARNING!

Start slowing down at about six or seven pumps so you will not injure your fingers against the piston if it rises too quickly.

You will likely find that it takes about $9\frac{1}{2}$ pumps to fill an empty taper.

Step 1 After loading, open the gate valve and turn the key counterclockwise until joint compound covers the leading edge of the tape. This is only necessary the first time you tape after each filling. Make sure the tape is loaded with the crease toward the gypsum board.

Step 2 Hold the taper with one hand on the control tube, called the slide, which is similar in operation to a shotgun pump. Put your other hand on the bottom of the mud tube so you can operate the creaser control lever. You might even put several fingers into the end of the mud tube if you find this gives you greater control.

Step 3 Start by taping ceiling butt joints, and then tape the ceiling flat joints. Use both drive wheels for the first 4" to 6" to secure the tape to the ceiling. Start at one end of the seam and work to the other end. After taping the first 6", tilt the taper at about a 20- or 30-degree angle away from the plane you are working in so that only one drive wheel is pressed to the board surface. This helps limit the amount of compound and prevents air bubbles. Walk backwards as rapidly as possible, leading with the head of the tool.

Step 4 As you approach the end of the joint, gradually bring the tool back to alignment with your vertical working plane. At about 3" to 4" from the end, stop and pull down sharply on the control tube, which will cut the tape. You will need to slow down and stop completely to do this because the tape or blade will jam if you do not stop to make this cut.

Step 5 Return the slide to its neutral position and bring back the other drive wheel so both wheels press against the surface once again. Keep both wheels rolling to maintain the continuous buttering of compound needed to press the last bit of tape onto the end of the joint. At this ending sequence (when both wheels are rolling), push the slide forward to eject the end of a new tape. It will be buttered with the correct amount of compound that allows you to start directly into the next run.

Using the Automatic Taper

Except for beginnings and ends of tape runs, always hold the automatic taper at about a 20- or 30-degree angle to the wall or ceiling and operate on one wheel only. It is important to have at least one wheel pressing against the board surface at all times while you are taping. These wheels control compound flow onto the tape. If you simply push the tape along without engaging a wheel, you will produce air pockets or bubbles, which are gaps in the tape where compound is missing. You should correct this immediately by going back, tearing out the unbuttered tape, and retaping. If you do not, the topping will dry over the air pocket and will eventually crack apart and crumble or fall off the surface completely.

NOTE

Be sure you do not continue using the automatic taper for more than ten minutes if no one is following behind you and bedding the tape.

Teamwork

On large finishing crews, one drywall mechanic operates the taper, another follows with a broad knife to wipe down excess compound, another spots fastener heads, another comes with the corner roller, and so on. If you are operating the automatic taping tool alone, however, you will need to put down the taper after taping a joint in order to wipe down excess compound.

Step 6 To finish the ceiling taping procedure, close the gate valve on the automatic taper and put it headfirst into a pail of clean water. Take a long-handled broad knife and wipe down all the ceiling butt and flat joints. The bedding joint compound should be wiped down while it is still wet, so that it is easily workable and excess moisture does not soak into the wallboard.

Step 7 Use firm pressure and hold your knife blade at about a 45-degree angle to the surface. Wipe along each taped joint, laying the tape flat and forcing out excess compound from underneath. Start in the middle of each seam and work first toward one end, then the other. This helps to avoid wrinkles and bunching up of the tape. Be sure to catch all the excess compound you squeeze out on your knife and scrape it off into a mud pan. The whole process is meant to embed the tape, fill the joint, and leave a generally flat surface.

2.3.5 Using the Automatic Taper on Wall Joints

After the ceiling is wiped down, the next process is to tape the horizontal and vertical wall joints. Again, start with the butt joints. Remember to open the gate valve on the taper.

Step 1 For vertical wall joints, place the automatic taper at the bottom of the joint, about parallel to the floor. Push the control tube forward one or two inches to make a tape leader. Start the vertical taping with the leader overlapping the floor a little. As you proceed upwards, the tape will be drawn up as you go. With a little practice, you will know exactly how long a leader to use so that the tape ends up exactly at the floor line after you have taped that joint. As soon as you can, when moving upwards off the floor, maneuver the taper so you are leading with the head. Also, shift so that you are tracking with just one wheel in contact with the wall.

Step 2 At 3" or 4" from the top of the wall, pull back the control tube to cut the tape and continue rolling to the ceiling intersection on both wheels. To start the next joint, roll the wheels slightly against the surface, starting the flow of compound while ejecting a new leader with the control tube.

Step 3 For taping horizontal joints, push forward on the control tube to produce a 2" or 3" tape leader. Place the leader, again with a slight overlap, at the beginning of the horizontal seam. Except for the start and end of a joint, always hold the taper at an angle to the wall (base of the tool angled downward) so only the bottom drive wheel is rolling against the surface. At the beginning of each seam, however, you need to push both wheels against the wall for about 6".

Step 4 As you come to within 3" or 4" of the end, pull back on the slide tube to cut the tape and continue on both drive wheels to the end of the joint while pushing forward on the slide. This applies the last several inches of tape while feeding out another leader for the next joint.

Step 5 For outside corners (where you are not using corner bead), simply follow the same procedure as explained above for vertical wall joints, but this time only apply tape to one side of the corner. Let the other side remain exposed to the air. When you have completed the vertical run, close the gate valve and fold the tape over the corner using your broad knife.

Step 6 At ceiling angle or wall intersection joints, you need to use the creaser wheel, which is extended by pulling on the trigger near the end of the automatic taper. You can also use the creaser wheel to help roll the tape against a flat seam, which is critically important when taping inside corners. Bisect the angle with your tape and make sure both wheels press equally on each side of the angle as you roll. Be sure to track in a straight line. Avoid twisting the automatic taper as you move. Again, start with a 1" or 2" leader to allow for the tape to be pulled toward the joint end (sometimes called "creeping"). You may have to push the leader into position in the angle using your fingers before you are able to proceed. Otherwise, taping inside corners and ceiling intersections is the same process as that used for vertical wall joints.

Step 7 If you are taping alone, stop using the taper after about 10 minutes, close the gate valve, and place the head into the pail of water. Then proceed to wipe down all the tape you just installed to embed it and remove the excess compound.

2.3.6 Using the Corner Roller

After placing the tape and joint compound in ceiling and corner angles, use the corner roller tool (*Figure 31*). This device embeds the tape in the joint compound at inside corners. It forces out excess compound from the tape. At the corners, you will need to remove excess compound with a broad knife. Then, after the angles are rolled, you need to go back over the full length of the angle with a corner finisher called a plow. The general sequence at all corners is as follows:

- Taping
- Rolling
- Plowing
- Corner applicator finishing

Four metal rollers in the head of the corner roller will embed and smooth the tape while forming a sharp corner crease. The tool is easy to use. Work it from the middle of the taped joint out toward the ends of the joint. This will force any overlap of tape due to stretching out to the end of the joint, where it can be trimmed off. This method stretches the tape in place. It also helps to prevent bunching up, which can easily happen if you start rolling at one end and go toward the middle instead of the other way around.

2.3.7 Corner Plow Operation

The corner finisher, or plow, is used to take out excess compound from angles after the corner roller has embedded the tape. This tool is available in widths of 2" and 3". You may use either size to wipe down the excess compound after the corner roller is used. Simply snap one of these tools onto the associated ball on the end of the corner applicator handle and use it like a plow.

With the arrow end leading, work from top to bottom for vertical angles, and from one end to the other for ceiling intersections. Wipe down these angles further on both sides with a 6" taping knife.

The plow is also used for topping, together with the corner applicator. It smooths and finishes both taped and topped corners. The corner finisher feathers the joint compound out from the corner and onto the drywall. It finishes both sides of the corner at once.

Another nickname for this tool is a butterfly, probably due to its shape. It has four skimming blades and is designed to wipe down and feather both sides of inside corners and other angles at once.

The plow has a spring action that compensates for corners slightly over or under 90 degrees. The blade design produces a smoothly feathered joint. When you use this tool, be sure the three tips, or arrows, are pointing in the forward direction of travel.

2.3.8 Flat Finisher Procedures

The flat finisher (box) adjusts the amount of topping compound applied to the joint. The compound is automatically feathered out from crown to edges. The crown runs down the center of the taped joint. This raised area is balanced out by the shrinkage that normally occurs when the compound dries in the joint. You can use a box to apply topping to both tapered and butt joints.

Under the box by the handle connection, you will find a small dial. This dial controls the size of the crown left by the box on the surface. It also controls the amount of topping applied to the surface. As you turn this dial, you will see the blade that controls the crown raise or lower. Establish the setting you will need for your particular application. In general, more crown is needed for earlier coats and less crown for later coats.

The flat finisher comes with handles of various sizes, usually from around 3' to 6'. Longer handles usually allow you to work at higher wall and ceiling levels right from the floor without needing stilts or scaffolds.

Always run the box with the wheels leading and the blade trailing. Adjust your grip in relation to your body so that you lead the box with the handle, except at the joint ends. Also note that the end of the handle has a special

Figure 31 Using a corner roller.
Source: Kevin Howser

A Different Type of Corner Finisher

With the MudRunner™ corner applicator, it is not necessary to apply pressure to the tool to feed compound to the plow. The operator merely turns the handle of the tool to dispense the compound stored in the reservoir just below the plow connection.

gripping lever. This grip locks the box in whatever position you desire (in relation to the handle) as you move across the surface.

To finish flat joints, proceed as follows:

Step 1   Load the box through its opening behind the blade using the adapter spout, or nozzle, on the loading pump. The box loads in the same way as the automatic nail spotter. In general, the topping compound should be a little thicker than the taping compound.

Step 2   Apply topping compound on wall or ceiling taped joints by drawing the box steadily along the joint while applying pressure to the back of the box with the handle. This forces out the compound evenly through the opening, depending on the crown you set on the dial. The blade also serves to feather the compound thinly out to the edges, leaving the crown in the center. Always start from one end of the seam and go straight across to the other end without stopping. It should always be a smooth process, and you should always make sure you have enough compound in the box before you begin the run.

Step 3   Ceiling joints should be topped first. The first ceiling joints to receive the topping are the butt joints. Butt joints receive a first topping coat, one coat on each side of the butt, using the 12" box on each side. Do not put this coating across the center of the butt; leave the center alone for the first coat. Set the dial on the 12" box to #1 (fullest) crown. Think of this as giving each butt joint a double-wide treatment. The reason for this is that you deliberately finish butt joints much wider than regular tapered (flat) seams to hide the very slight crown already caused by the butted wallboard panels. The finished crown is so gradual and so slight that it will hardly be noticeable.

Step 4   The flat joints receive topping after the butt joints. Again, start with the ceiling. Use the 7" flat finisher set at #3 (medium) crown. Start at one end of the joint and apply even pressure to the middle of the joint. Lock onto the grip as soon as possible after starting the run. Lead with the handle.

Step 5   When you approach the middle of the run, keep the lock on the handle and gradually release the pressure. Then remove the box in a sweeping motion from the surface.

Step 6   Next, reverse hand positions and start at the other end of the run. Repeat the process described above for beginning the run. Lead with the handle toward the middle of the joint where you stopped before. Again, keeping the lock on, slightly overlap the stopping point and remove the box with a sweeping motion.

2.3.9 Flat Finishing Other Joints

For wall vertical flat joints, start at the bottom and lock onto the handle grip right away; then remove the box in a sweeping motion about 2' to 3' above the floor line. Start at the top of the joint and apply pressure down to your previous stopping point. Again, with the handle locked, slightly overlap that point and then sweep off the surface, which should neatly finish the two topped sections.

When applying topping to joints near doors, windows, and other openings, always work from the corner and move towards the opening. Just before the wheels reach the opening, keep the handle locked and lift the wheels. Then sweep away out into the opening so that topping compound is applied all the way to the edge.

2.3.10 Finishing Coats with the Flat Finisher

Second and third topping coats are applied in the same way as the first topping coat, but in thinner and wider layers. These finishing coats are applied to fill out minor shrinkage and unevenness that could produce shadows and other imperfections after painting. Always allow each coat to dry overnight. In cold or humid conditions, each coat might take even longer to dry.

Before starting each finishing coat, scuff the surfaces. This means lightly sanding the dried compound areas to remove any crumbs, burrs, globs, and so forth. This also prevents any of these things from producing scratches in the surface when you make another pass over them with a box. Be sure to wipe down all sanded areas.

Fill the 10" box, set the dial to #3 (medium), and then apply a straddle coat on all butt joints, starting with the ceilings. Apply this coat of topping compound right down the center (on top of the taped seam) between each of the previous double-wide coats. A correctly finished butt joint should always have a final width of at least 25".

Do the flat seams next. Use the 12" box if only one coat is required, or use the 10" box for the second coat, and the 12" box for the third. Reset the dial to #5 (least) crown and cover the ceiling and wall flat joints in the same way as for the first topping coat.

2.3.11 Corner Applicator Operation

In topping operations, the corner applicator provides a final finish for ceiling and inside corner angles. Use it after the bedding coat has dried. Attach the corner finisher to the corner applicator by snapping it in place. The plow becomes the skimming or troweling blade for the tool. This runs in the topping coat at the corner angles. The corner applicator smooths and feathers the final coat. The corner applicator does for angles and corners what the flat finisher does for flat and butt joints.

To use a corner applicator, proceed as follows:

Step 1   To load the corner applicator, first remove the nozzle end from the chain-slung filler adapter that goes on the loading pump. Inside the nozzle housing is a rubber O-ring that prevents leakage during filling. The filler valve on the corner applicator is inserted into the housing against the O-ring, which seals it. Pressure from the pumped compound pushes the tool's filler valve open.

Step 2   Once the corner applicator is full of topping compound, attach the 3" plow over the round opening. Place the tool at one end of the corner angle. Then, with the nose of the plow leading, draw the tool along the angle, applying steady pressure with the handle on the back of the box. This forces the compound out where the plow distributes and feathers it along the run.

Step 3   As you near the end of the run, sweep the tool away from the surface in the same way as with a flat finisher. Reverse hand and body positions and start again at the unfinished end. Draw the tool back to the previous stopping point. Overlap just a little and then sweep away from the surface. This should neatly join both sections of the run. Make sure you apply the plow to neatly bisect the angle. Also, keep the tool at as close to a right angle to the corner as possible.

Step 4   To apply compound to vertical angles, start at the top of the wall and draw the tool downward, sweeping away from the surface at about knee height. Then place the head in the bottom of the angle seam and draw the tool back upward to barely overlap the previous stopping point. Gradually sweep away from the surface, neatly joining both sections of the run.

Step 5   Detail out the corner intersections with a broad knife. Feather any excess compound away from the angle on both sides, also with a broad knife.

When the plow is replaced by a ribbon dispenser attachment, the corner applicator becomes a tool known as a *flat applicator*. The flat applicator is an alternative tool to the automatic taper. It applies a bedding coat of taping compound and allows you to then attach tape to the surface by hand.

This semiautomatic taping method is useful in places where you physically cannot work with an automatic taper. There is also a mini-taper that might work just as well in confined areas. The flat applicator is convenient for emergencies, for hand operations, or for use by one mechanic when another mechanic is using the only taper on the job.

All these automatic taping and topping tools must be kept clean. Keep the tool heads submerged in a bucket of clean water whenever they are not being used.

2.4.0 | Hand Finishing Procedures

Drywall finishing procedures start with gathering all the proper tools, equipment, and materials necessary to do the job. Decisions are then made about sequencing the various tasks: what is done first, second, and so on. These tasks include the following:

- Inspecting, repairing, and prefilling
- Taping flat joints, corners, and other angles
- Installing bead and trim pieces
- Spotting fastener heads
- Topping, scuffing, and sanding

To finish drywall, proceed as follows:

Step 1 Countersink the nail and screw heads and cover them with joint compound. Damaged drywall must be patched. Do only a minimum of scuffing or sanding on the paper so as not to roughen it or raise the nap.

Step 2 Once the drywall has been properly installed, carefully inspected, and fixed where necessary, the next step is to prepare for spotting and taping. The usual finishing sequence is to spot the fastener heads; pre-fill gaps, damaged areas, and butt joints; tape the ceiling joints; and then tape the corners, other angles, and finally the flat joints.

Step 3 Prepare the joint compound according to directions and the jobsite conditions. Put down a suitable sheet of scrap wallboard first. Do all your mixing at this one place on top of the protective scrap board.

Step 4 For hand taping, load the mixed compound into the bladed mud pan. Obtain the proper compound consistency. It should generally be the consistency of soft putty.

Step 5 Apply the compound to the bare joint with a broad knife. While the joint compound is still wet, apply the joint reinforcing tape, making sure the correct side of the tape is against the drywall. Press the tape into the compound (*Figure 32*). You can use a broad knife to smooth the tape. Force the excess compound out from under the tape and remove it with the knife. This bonds the tape to the compound. For perforated tape, force the excess compound up and out through the holes. Again, wipe away the excess with the knife. Make sure there is enough joint compound under the tape or bubbles will result.

Step 6 Spread a thin coat of joint compound over the top of the tape. Allow these bedding coats of joint compound to dry overnight. Special precautions need to be taken when finishing butt joints. For hand finishing, it is critically important to look at the taped butt before you finish it. Follow these guidelines when finishing butt joints:

- If the butt joint has a high side and a low side, coat the low side.
- If the wallboard butt forms a hollow, fill it back to the normal plane by adding much more compound than usual.
- If the butt joint is regular, it still needs to be finished with the double-wide method (*Figure 33*). For the first topping coat, do not cover the tape. On day

Figure 32 Applying tape.
Source: Andrey tiyk/Shutterstock

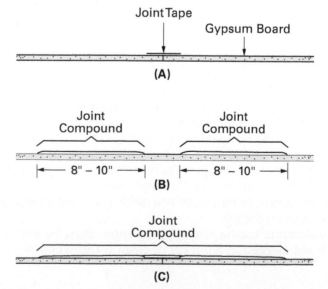

Figure 33 Finishing a butt joint using the double-wide method.

one, make a crown of compound on each side of the tape. Then, on day two, apply the straddle coat to cover the tape between the crowns, thereby making one slightly larger crown in the middle of the seam.

Step 7 Apply a thin coat of joint topping compound. Feather the compound's outer edges. Feathering spreads the compound thinly from the center of the taped joint outwards beyond each edge of the tape, causing a slight crown over the center. However, the smaller the crown and the finer the feather, the less sanding will be required.

Step 8 Allow this coat to dry. Some compounds require 24 hours to dry; some take even longer. Drying times also depend upon atmospheric conditions within the structure. If the building is only partially closed off, the finishing work will be affected by the outside weather. If it is too cold or wet, the compound might not dry out at all until the weather changes. Finishing work depends upon a controlled interior.

Step 9 Once the coat has dried thoroughly, sand it smooth. Remove the sanding dust. Depending on the situation, you may be applying another topping coat, or one or more skim coats. These determinations are almost always made by your supervisor.

2.4.1 Sanding

Always wipe down after any light or heavy sanding to remove sanding dust and tiny particles of compound or other debris. Always check the coated surface to see if it is straight and smooth. Use a straightedge or level to do this.

Always have mud pan available filled with topping compound, no matter what finishing procedure you are doing. This way, whenever you see something that needs a little touch-up, like a pit or scratch, you can take care of it immediately to avoid poor-quality finishing.

For general sanding and scuffing procedures, 100-grit, 120-grit, or 150-grit sandpaper is recommended. Sand screen is also used for scuffing as well as for final sanding, because it does not raise the nap on drywall face paper. Sandpaper of less than 100 grit should be avoided, and any sandpaper coarser than 80 grit is unacceptable.

Remember, you are sanding a dried, porous wall joint covering material in order to smooth out tiny bumps and spaces so it will hold a final decoration as well as wallboard face paper. Guard against oversanding, which tends to grind out hollows.

> **WARNING!**
>
> Be sure to wear eye protection and appropriate respiratory equipment when sanding. Check the SDS for the applicable drywall to learn about the safety hazards associated with that material.

Correcting Oversanding

Excessive sanding or use of coarse sandpaper can cause the paper fibers of the drywall to stand up. If the problem is not too severe, light sanding with a very fine sandpaper or wiping the panel down with a damp sponge or cloth can correct it. Otherwise, use a light skim coat of topping or all-purpose compound to correct the problem.

2.4.2 Spotting Fastener Heads

Check the drywall screws. Be sure they have been dimpled (set below the substrate with a hammer). Apply the first coat of joint compound on top of the screw heads. Allow it to dry. Sand the dried coat with an abrasive cloth or sand screen. Apply a second coat of joint compound. Allow this coat to dry. Sand it smooth and apply a third coat. The covered area should be smooth and level with the substrate.

Many contractors prefer hand spotting and will not allow the use of a nail spotter tool for this task. Their reasoning is based on the following factors:

- Fasteners may not always be below the substrate level, so the nail spotter's blade is frequently damaged. This causes downtime to change blades or tools. When there are no more replacement blades or tools, hand spotting will be the only remaining option.
- Spotting by hand is a reasonably fast method when done by an experienced mechanic.

- Hand spotting is thorough. It forces the compound more effectively onto the head and completely fills the dimple. It also packs compound down into the crisscross of Phillips-head screws better than the automatic tool usually does.

- Hand spotting is something even the newest apprentice can master almost at once. It helps technicians appreciate the nature of finishing work faster than any other process. It can also build confidence, speed, and an eye for detail.

All fastener heads need to receive three coats of topping, so they are undetectable when the surface is finished. Use a 5" knife for the first two coats and a 6" or 8" knife for the third coat. Make sure the compound fills in the crossed indentations in Phillips-head drywall screws.

2.4.3 Outside Corners

To finish outside corners, proceed as follows:

Step 1   Attach the metal corner beads to the outside corner angles. Fasten them with drywall nails or a clinch-on tool or by applying tape with compound. Staples are sometimes used, although many contractors try to avoid them. Dimple the nail heads. Apply joint compound to each flange with a broad knife right after fastening.

Step 2   Spread the compound 7" from either side of the nose (center outside corner). Cover the metal edges with compound. Allow the compound to dry. Sand it lightly, then apply the next coat. Feather the coat out two inches from the first coat. Let this dry and sand it smooth. Also apply and smooth a third coat of topping at outside corners. Use an 8" knife for all three coats, a 6" knife for the first coat and an 8" knife for the others, or an 8" knife for the first two coats and a 10" knife for the third coat.

The goal with each coat is to fill the corner so it is flat, not concave. Too much compound will make the corner concave. Finish each outside corner so that the corner bead is completely invisible.

2.4.4 Inside Corners

To finish inside corners, proceed as follows:

Step 1   Cut the joint tape to the length of the corner angle. Apply joint compound to each side of the tape angles. Apply small amounts of compound to both sides of the corner angle. This prevents thin cracks from occurring at the angle. Crease the tape length along the center.

Step 2   Use a 4" broad knife to press and embed the tape in the compound. Apply enough pressure to wipe the compound from under the tape. Feather the compound 2" beyond the tape edges. Let it dry and then sand. A corner tool is available as an alternative (*Figure 34*).

Step 3   Apply and feather the next coat about 2" beyond the first coat. Dry and sand. Then apply a third coat where needed.

Step 4   For inside corners, it is better when hand finishing to apply compound first to one side, wait overnight, and then apply compound to the other side. With the automatic corner applicator/finisher tool, you only need apply one coat of topping to inside corners. This is called plowing or glazing the angles.

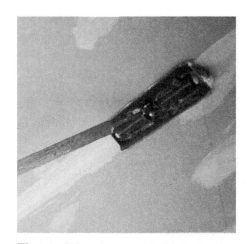

Figure 34 Using a corner taping tool.
Source: Tetra Images/Alamy Stock Photo

2.4.5 Safety and Good Housekeeping

Follow all recommended safety practices for drywall finishing. Maintain your tools and equipment. If you use stilts instead of a ladder, practice safety; always put stilts on and take them off when you are able to lean against a wall, preferably in a corner. Never climb stairs or try to pick up something off the floor while wearing stilts. Ask for help from someone who is not wearing stilts.

Be aware that taping and topping tools, knives, and trowels carry with them a certain degree of hazard. For example, even a dull blade can cut or injure an eye or face. Any tool, if mishandled, can cause an accident or injury. If tools are allowed to clutter up a work area, they can also cause an accident or injury. Keep unused tools in a safe place out of everyone's way.

One of the most common hazards when finishing is slippery conditions, often caused by wet compound carelessly spilled on the floor. This is especially hazardous for someone on stilts. The best rule is, if you spill something, clean it up—no matter what it is. If wearing stilts, either remove them and clean up the spill or ask someone who is not on stilts to clean it up for you; do not attempt to bend over on stilts. The same is true if you drop a tool.

Store finishing materials in a cool, dry, protected location. Provide adequate ventilation during dry sanding. Always wear a face mask or respirator to prevent inhalation of sanding dust. Wash hands after applying joint compound as well as after sanding. Proper safety and housekeeping procedures minimize illness and injury.

Final cleanup is always your responsibility. This means a complete sanding and wipe-down of all ceilings and floors, and a thorough scraping and sweeping of all floors.

Keep in mind that sanding dust travels. If you are doing remodeling or repair in a finished building, secure the area in which you are working by covering doors and other wall openings with plastic. If you are working in a room that contains furniture, equipment, or other items, cover them.

Using Stilts

Do not try to use stilts unless you have been properly trained to use them: If you plan to use stilts, make sure they are not prohibited by local codes.

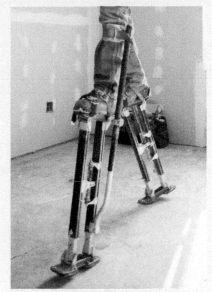

Source: Tetra Images/Alamy Stock Photo

2.0.0 Section Review

1. During winter conditions, drywall finishing should not be attempted unless the building has heat in a somewhat controllable range between _____.
   a. 70°F and 90°F
   b. 40°F and 60°F
   c. 30°F and 80°F
   d. 50°F and 80°F

2. The first step in the drywall finishing process is to _____.
   a. complete repairs, cutouts, and prefill
   b. sand and scuff angles and flats
   c. apply topping coat over taped joints and seams
   d. complete pole and hand sanding

3. A(n) _____ embeds the tape in the joint compound at the inside corners.
   a. automatic taping tool
   b. corner roller tool
   c. corner plow
   d. flat finisher

4. Sandpaper coarser than _____ is unacceptable to use during drywall finishing.
   a. 150 grit
   b. 80 grit
   c. 120 grit
   d. 100 grit

	# 3.0.0 Basic Repair
Performance Task	**Objective**
4. Patch damaged drywall using two different methods.	Explain the importance of knowing how to repair damage, list types of damage, and describe steps to fix different kinds of damage.

    a. Describe finished joint repair.
    b. Describe repair procedures for fixing problems with compound.
    c. Describe fastener-related damage and how to repair it.
    d. List problems with wallboard and explain how to fix them.

The true test of your finished work will come not from how well you avoid making mistakes, but in how well you repair your mistakes. You may also have to repair those problems left for you by someone else. This section explains the areas where the most common problems are likely to occur and gives you information on their causes and how you can fix them. The four main problem areas are as follows:

- Joints
- Compounds
- Fasteners
- Gypsum board panels

3.1.0 Finished Joint Problems

The common joint problems found in drywall work include the following:

- Ridging
- Tape photographing/telegraphing
- Joint depressions
- High joints
- Discoloration
- Tape blisters
- Cracks in the joint

These joint problems are described in detail in the paragraphs that follow.

3.1.1 Ridging

When a ridge occurs along a joint between two boards, it is often because there has been movement at the joint. Ridging is also sometimes known as beading or picture framing because a visible ridge that surrounds a board resembles a frame surrounding a picture. The following are three probable causes of ridging:

- *High humidity, poor heat distribution, or not enough ventilation* — This results in expansion and contraction of the framing and boards.

- *Drywall that is not properly installed* — Improper installation includes misaligned framing and butt joints that do not fit well together. If you force two boards together, joint compound may be squeezed out, forming a ridge. Make sure the joint is not overstressed by too tight a fit.

- *Too much joint compound* — To correct ridging caused by too much joint compound, first sand the ridge smooth, then apply a finishing coat of joint compound. Hold a light at an angle to the area to make sure you have eliminated the ridge and left a smooth surface.

Onsite

Improperly Installed Drywall

If ridging occurs along a joint between two boards, the drywall may not have been installed properly.

You may have to cut some of the drywall away at the joint to make a little gap between the boards. Do this by cutting with a knife, making several passes, or using a hand saw or chisel to remove a sliver of one of the boards. If the space left between the boards is too wide, the joint will be weaker than a properly fitted joint. Ideally, the entire width of the joint tape is bonded to a drywall surface. The tape itself only needs to span a small space between the boards. If the space between the boards is too wide, less of the tape is bonded to the drywall. This can weaken the joint and promote ridging.

You may have to overlap two pieces of tape to strengthen the joint. If the joint is very wide, it may be because one board, or both, is not securely fastened to the framing at the joint. Perhaps neither board is directly over the framing member. You may have to install fasteners at an angle through the boards into the framing to make the joint tight against the framing.

This presents a problem; one of the most basic rules is always to drive the fasteners straight so that the edges of the heads will not protrude. In this case, however, they will. Repair the problem by using a hammer to straighten the head (by bending the shank) after the fastener is installed. You may also need to cut and install a thin strip of gypsum board to bridge a space that is too wide.

3.1.2 Photographing

If the joint is still visible even after the wall is finished and painted, this condition is known as *photographing*, also known as *joint telegraphing*. The joint may show through as a slightly different color than the finished wall. Or it may be the same color as the finished wall but have a higher or lower gloss (shine) to it. Photographing can also occur over fasteners if there was insufficient joint or topping compound spotted on the heads.

The usual causes of joint photographing are the following:

- The installer failed to force excess joint compound out from under the joint.
- High humidity conditions delayed drying of the second and third topping coats of compound.
- The tape absorbed too much moisture from the compound, causing the joint compound to shrink and conform to the shape of the joint tape. Avoid this by wetting the tape before installation.

To correct photographing, sand the tape edges to feather them into the surface of the wall or ceiling. Then cover the tape with thin coats of joint or topping compound. Use thin coats so as not to rewet the tape too much, or seal the tape with a primer after sanding and before applying the final coats of joint compound. This keeps the tape from drawing too much moisture out of the joint compound.

3.1.3 Joint Depressions

A joint depression is a valley that occurs at a seam or joint. It will be most obvious when a light strikes the drywall at an angle. (Hint: Use a 10" or 12" knife to help find high and low spots.) Two common causes of joint depressions are the following:

- There may not be enough joint compound over the joint. This can happen when the joint compound mixture is too thin or when not enough joint compound is applied to the joint.
- The joint may be sanded too deeply.

The cure for joint depressions is to add more joint or topping compound to the joint. Then smooth it and sand again to get a flat surface at the joint. Make sure the joint is flush with the surface of the wall or ceiling.

Figure 35 Coating with topping compound.
Source: New Africa/Shutterstock

3.1.4 High Joints

A high joint is the opposite of a joint depression. It occurs when a wide section of joint is raised above the rest of the drywall surface. Like the joint depression, a high joint is most noticeable when a light strikes it from an angle. High joints are the result of too much joint compound built up underneath or on top of the tape and/or improper feathering of each coat. The edge of the coat must be feathered into the wall-board surface. When this is not done, high joints will result.

To repair a high joint, sand the area as flush as possible without sanding into the tape. Then apply one or two final skim coats of topping compound. Feather each coat onto the board surface. Make each coat wider than the previous coat to conceal the area of the joint (*Figure 35*).

3.1.5 Discoloration

Joints may discolor or turn lighter or darker than the rest of the finished surface. The following are several common reasons for joint discoloration:

- Moisture may be trapped inside the joint. Until a joint is sealed, it can absorb water. If a joint is sealed before it is dry, water is sealed inside the joint. Trapped water will degrade the finish and discolor the surface. Be sure the joint is dry before sealing it.

- The joint was painted in conditions of excess humidity. To prevent this, reduce the room humidity before any painting is done.

- A poor-quality paint was used. Always be sure to use a good-quality paint. Cheap paint often gives uneven coverage and sealing. This increases discoloration.

3.1.6 Tape Blisters

A tape blister is really an air bubble in the surface of a joint. It can be several inches long or as small as a dime. Tape blisters occur when the bond fails between the tape and the first bedding coat of joint compound. There may be no bond to start with if care is not taken when using an automatic taper and sections of tape come out without a joint compound coat underneath.

Tape blisters can also happen if the joint is too wide, either because the tape was not properly embedded in the joint compound, or because the tape draws moisture too quickly from the joint compound. Another cause of a blister occurs when topping compound is used instead of joint compound to embed the tape.

To repair a tape blister, proceed as follows:

Step 1   Slit the blister with a knife. If the blister is large, cut and remove the entire section of tape that came unbonded.

Step 2   Sand or scrape out enough of the dried joint compound so you can embed a new section of tape.

Step 3   Work joint compound underneath the tape, smoothing the slit in the old or new section of tape into the joint compound as you go. This embeds the blistered section. This is a hand procedure only. Do not attempt to do this with another run of the automatic taper.

Step 4   Apply a skim coat of joint compound over the tape. When this coat is dry, apply the required number of topping coats, always allowing enough drying time in between coats. Sand enough to produce a smooth finish that is flush with the surrounding surface.

3.1.7 Cracking

The two common types of joint cracks are those that run along the edges of a joint and those that run along the center of a joint. Each has its own causes.

Edge cracks are cracks along joint edges that occur when the air temperature is high and the humidity is low when the joints are finished. This causes the joint compound to dry too quickly and unevenly, resulting in uneven shrinkage.

To slow down the drying rate, run a wet roller over the joint or spray it with a fine water mist from an atomizer.

Edge cracks can also be caused by tape that has a thick edge or by joint compounds applied in coats that are too thick.

The procedure for repairing edge cracks depends on whether the crack is thin or wide. If the crack is small and thin, coat it with a latex emulsion or a thin coat of joint or topping compound. Then sand it as needed.

If the crack is wide, you may have to gouge out some of the joint compound to prepare the surface. Paint the gouged-out crack with a primer, then fill with joint compound and sand smooth.

Center cracks are cracks running along the center of a joint that occur for the following reasons:

- If the tape is still intact, the crack is probably the result of applying joint compound too thickly. Also, low humidity may have caused the joint compound to dry too quickly.

- If the tape under the crack has been torn, it is possible that the structure is settling or another type of movement caused the crack.

To repair cracks along the center of the joint, follow these procedures:

- If the tape is still intact and the crack is narrow, apply spackling compound to the crack. If the crack is wide, use joint or topping compound to bridge the space.

- If the tape is torn, you may need to remove a section of tape and old joint compound before making repairs. Then retape the joint following the usual finishing procedures.

3.2.0   Problems with Compound

Compound has its own special set of problems. It can debond, grow mold, become pitted, sag, and shrink.

3.2.1 Debonding, Flaking, or Chipping

When the joint or topping compound will not bond to the tape or the board or becomes unbonded from either one, the condition is known as compound debonding, flaking, or chipping. Common causes of compound debonding include the following:

- A foreign substance was on the gypsum drywall surface or on the surface of the tape when the joint compound was applied. Examples of foreign substances include dirt, oil, sanding dust, and incompatible paint.

- The joint compound was mixed improperly, or the wrong ratio of water to dry powder was used to mix the compound.

- Too much water was added during mixing, or incompatible compounds were mixed with each other in order to add to the working supply, combine containers for storage, etc.

- Dirty water was used to mix the compound, or dirty tools were used to mix or apply it.

- Hot or heated water was used to mix the compound. One reason for avoiding hot water is because of possible sediment problems associated with hot water heaters.

- The installer used old or expired compound.

You can avoid most compound debonding problems by following the manufacturer's mixing and usage instructions exactly. Some manufacturers request that you let the compound sit for a while after mixing it. There is a good reason for this, so do not take any shortcuts. They will end up costing excessive time and waste in the long run.

Be sure to always use only clean, cold water to mix any compound. Also use clean mixing and application tools and equipment. Remember that automatic taping tools need to be kept in pails of clean water between uses. Always make sure the gypsum drywall surface, tape, and all mixing pails are clean, too.

Repairing compound debonding is very much like repairing tape blisters, only on a larger scale. First, separate the debonded section of tape from the dried joint compound. Then remove enough of the old joint compound to allow you to apply a new layer in which to embed the tape. If the old compound crumbles easily, remove all of it. You will also have to remove whatever joint compound was used to feather the joint. Apply new compound and tape as you would for a new joint.

3.2.2 Moldy or Contaminated Compound

Moldy compound smells terrible, so you will have no trouble recognizing it. Mixing with contaminated water, using dirty containers or tools, or letting the compound stand too long can result in mold, bacteria, and bad odors in the compound. Hot and humid weather also contributes to the growth of mold and bacteria.

Always be careful to examine every pail before mixing anything in it. You may be surprised at what you might find in a supposedly empty bucket. The best remedy here is simply to look before you mix. If you discover that your batch of compound has become moldy or contaminated, throw it out. Then be sure to soak your tools and containers in a solution of chlorine bleach and clean water at least overnight.

Be sure to clean and wash all tools and equipment components at the end of every working day. The joint compound intended for use the next day must be stored in covered containers and kept at room temperature overnight. That means warm room temperatures, not freezing cold or scorching hot.

3.2.3 Pitting

Small pits may appear in the finish of the joint compound after it dries. Pitting has the following common causes:

- Air escaped after being trapped in the joint compound mixture. This can happen if you mix the compound too vigorously or for too long.
- The joint compound mixture was too thin.
- Not enough pressure was used to apply the joint compound to the joint; that is, it was not embedded or wiped down properly.
- Joint compound was not adequately mixed prior to application.

To prevent pitting, mix the compound thoroughly using a slow, steady motion. Set power mixers, if used, on slow speed. You are trying to create a smooth mixture that is free of lumps. When you apply the joint compound, use enough force to establish a good bond to the surface, smooth it out, and feather the edges.

Repair a section of pitted compound by simply skim-coating with another topping layer to fill the pits. You may need to sand a little to form a smooth base for applying the new joint compound. Then apply the new compound as you would apply a topping coat to the joint. Feather it out to conceal the joint area. You may have to feather it wider than the original topping coat to completely hide the joint.

3.2.4 Sagging

When compound sags or shows evidence of runs, the following conditions are usually present:

- The joint compound was too thin. When mixed properly, compound is thick and smooth. Be sure to follow the mixing instructions exactly.
- Water added to the joint compound or to the dry powder compound was too cold to mix completely. Again, cold water is essential, but do not use ice cold water. Remember that finishing is a room-temperature process. Anything colder than what normally comes out of a faucet in a warm room is just too cold.

To repair sags and runs, sand them very smooth after they dry. Then, recoat with layers of joint or topping compound as needed.

3.2.5 Excessive Shrinkage

Joint compound that shrinks too much when it dries is probably the result of one of the following:

- Joint compound mixed too thin
- Insufficient drying time between coats
- Too much joint compound applied at one time

To prevent this problem, use lightweight joint compound, which tends to shrink less. This problem is similar to the joint depression problem. As in that case, remedy excessive compound shrinkage by applying more joint compound. However, ensure that each previous coat is thoroughly dry before you begin any repair by adding more compound.

3.2.6 Delayed Shrinkage

Delayed shrinkage is caused when too much time elapses before the correct amount of shrinkage occurs. The joint compound is not shrinking enough and tends to resist drying out. Delayed shrinkage has the following common causes:

- Atmospheric conditions (slow drying capabilities and very high humidity)
- Insufficient drying time between coats of compound (trying to rush the job before it is ready for each finishing procedure)
- Excess water added to the joint compound mixture
- Heavy fills (adding too much joint compound as a prefill or trying to fill large spaces in the gypsum drywall with compound instead of slivers or strips of wallboard)

One way to prevent delayed shrinkage is to use a faster-drying compound, perhaps a quick setting compound, which sets up chemically and does not depend on water evaporation. Quick setting compounds were previously discussed in this module.

A remedy for this condition is to allow extra drying time and then to reapply a full cover coat of a heavy-mixed joint compound over the tape. Most shrinkage will generally take place on this heavy topping coat. With the right joint compound, the coat will dry faster and allow you to continue finishing procedures in the usual way.

The best defense against delayed shrinkage is to use a faster-drying compound in the first place. There is very little you can do to joint compound that needs more drying time, except to give it more time to dry.

3.3.0 Fastener Problems

Two common fastener problems that may be encountered are nail pops and fastener depressions. These problems are described in more detail in the paragraphs that follow.

3.3.1 Nail Pops

When drywall fasteners work up from under the finished surface after the installation is complete, the job is said to have nail pops. Whether nails or screws are used as fasteners, this is still referred to as a nail pop. Nail pops are unsightly, protruding fastener heads.

If enough fasteners pop out, the drywall will loosen and sag. The fasteners can be driven in again and the hole refinished, but the best remedy is preventing nail pops before they happen. Following are the primary reasons for nail pops:

- Wood framing with relatively high moisture content will shrink as the lumber dries out. As the wood shrinks, the fasteners lose their tight holding power (*Figure 36*). When the wallboard is no longer securely attached, a space develops between the board and the stud; the fastener shank is exposed at that point. Then almost anything that puts pressure against the wallboard will push it against the stud. The fastener—which does not move—will actually pop right out of the panel along with the compound covering it.

- When drywall is fastened to framing that is out of alignment, stress on the drywall causes fasteners to work up above the surface (*Figure 37*).
- Gravity acting on ceilings and vibrations acting on walls will tend to work the fasteners loose (*Figure 38*).
- The drywall may not have been installed properly (*Figure 39*).
- If a building has poor ventilation or an inadequate heating system, large temperature fluctuations will cause expansion and contraction of the framing and drywall. If there is too much of that, the fasteners will begin to loosen.

Nail Pops

Nail pops that occur after the building has been heated for more than a month are usually caused by lumber shrinkage. Once this begins to occur, it is better to wait until the end of the heating season to repair the nail pops.

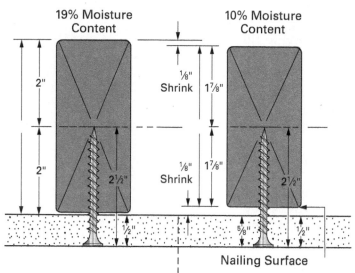

Figure 36 Shrinkage contributes to nail pops.

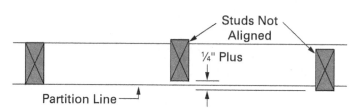

Figure 37 Non-aligned framing.

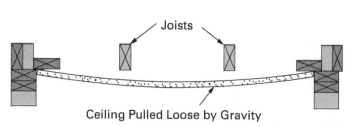

Figure 38 Force of gravity can cause nail pops.

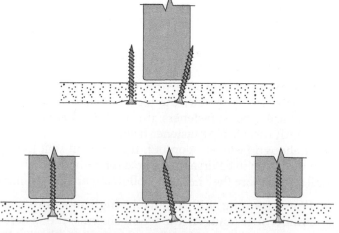

Figure 39 Improper drywall installation.

3.3.2 Preventing Nail Pops

Nail pops may show up days or weeks after installation is complete or gradually over a period of many months. How soon they appear depends on the degree of misalignment, the type of fasteners used, the moisture content of the framing at the time of installation, the amount of vibration present, and the temperature cycles to which the drywall and framing are exposed. Once you know the causes of nail pops, prevention is easier. To prevent nail pops, follow these key rules:

- Make sure your framing lumber is dry before you fasten any drywall to it. The builder should provide enough ventilation to speed up the drying process. When working in cold or humid weather, use portable heaters or blowers to warm or circulate the air. It may take only a few days to reduce the moisture content of lumber to an acceptable level, depending on temperature and humidity. Lumber is too green for hanging drywall if any wet spot appears when the lumber is hit sharply with the head of a hammer. Test several lengths before forming an opinion. The amount of moisture can vary from one piece to the next.

- Make sure the framing members are aligned in the same plane. Sight along the edges of the studs and joists to see that they are in a straight line. You can also check alignment by holding a long straightedge up against the studs and joists.

If framing is out of alignment, repair it. Applying boards to framing that is out of alignment will prove to be a mistake. The framing will eventually spring back to its original position and the nails will pull out from the framing.

- Use appropriate fasteners to attach the drywall boards to the framing. Use floating corner angles to reduce stress on the drywall. Finally, you can use adhesive in addition to nails or screws to fasten the drywall boards.

- Always work from the center of the drywall board toward the edges. If you work from one edge of the board to the other, the board may not rest firmly against the framing for the entire length of the board.

For example, assume that you are installing the final board in a wall. The other boards along the wall are already in, and so are the boards that cover the wall that forms the other half of the corner. On this final board, if you start installing the fasteners at one edge instead of at the center, the drywall may move slightly toward the opposite edge. This particular board is trapped by a corner and other drywall boards. Any movement after you start nailing will stress the board, eventually causing it to bow and pop the fasteners.

A similar problem can occur if you work from the other edge toward the center of the board. The center may actually spring away from the framing. If you fasten the center of the board first and work toward the edges, the board will not be able to move once the first fasteners are in place. From the moment the first fastener is driven home, the board is forced flat against the framing in the center and along the edges. Be sure to hold the drywall board tightly against the framing as you install the fasteners. This ensures that the board stays flat. Do not worry about installing a board with a slight bow. With proper installation, any stress put on the board by flattening its bow will relax in a short time.

Before you cover the fastener heads on any drywall finishing job, check to be sure they are tight. Re-drive any loose fasteners. It is also a good idea to drive another fastener on each side of a fastener that has worked loose. Drive them about $1\frac{1}{2}$" away from the old fastener. After you re-drive any loose fasteners and add extra fasteners, go back and check all the fastener heads again. The vibration may have loosened more fasteners. The few seconds you spend now can save you a few hours later. Also, if you fastened boards to both sides of a wall, driving the fasteners on one side may have loosened those on the other side. Be sure to check the first side again.

Attention to detail should prevent most nail pops. Take the time to check your framing lumber carefully and install your panels properly.

3.3.3 Fastener Depressions

A fastener depression is a depressed area over the fastener head. This is the opposite of a nail pop. The joint compound over a nail or screw has sunk lower than the surface of the surrounding drywall.

Fastener depressions can be caused by the following problems:

- Nails were depressed too deeply or screw heads driven in too far.
- Not enough joint compound was applied to the fastener heads to cover them properly.
- The framing lumber was extremely dry. Dry lumber will absorb moisture, squeezing the board between the nail head and the edge of the stud or joist and pulling the fastener head deeper into the drywall.
- The installer used too few fasteners to hold the drywall firmly against the framing, allowing the drywall to flex independently of the framing and forcing the fastener heads deeper into the surface.

To prevent fastener depressions, avoid driving fasteners through the facing paper. Install the correct number of fasteners and space them properly. Spot the fastener heads with three coats of compound, sanding lightly between coats, if necessary.

Repairing fastener depressions is a simple matter. First, make sure you have installed enough fasteners. If you need more nails or screws to hold the drywall firmly against the framing, add them. Second, spot the fastener heads with joint compound to bring the surface flush with the surrounding drywall.

3.4.0 Wallboard Problems

Common problems with gypsum drywall sheets include blisters, damaged edges, water damage, board bowing, board cracks, fractures, and brittleness.

3.4.1 Board Blisters

When the facing paper becomes unbonded from the surface of a piece of gypsum board, it is known as a board blister. It may be caused by a manufacturing defect, or it may be the result of careless handling or improper storage. The gypsum filler tends to break apart inside the wrapped board, causing the facing paper to loosen.

There are two common ways to repair board blisters:

- Inject an aliphatic resin glue, such as yellow or white carpenter's or wood glue, into the blister, and then press the paper flat. This is the best remedy where the blister is small or where the blister is not discovered until after the wall has been textured and/or painted.

WARNING!

Before using any adhesive, check the manufacturer's instructions and applicable SDS to identify any hazards. Wear protective equipment and apparel as specified by the manufacturer.

- Cut out the entire blistered area and finish it with tape and joint compound. Follow the usual procedure for embedding tape and finishing joints. If one width of tape is not going to be enough to cover the blistered area, add as many other strips as necessary.

3.4.2 Damaged Edges

Improper handling of gypsum drywall sheets is what generally causes damaged surfaces and edges. Such carelessness may cause the facing paper to tear or the gypsum core to crumble.

The only way to repair such damage is to cut off the damaged area back to sound gypsum board prior to installation.

If a board has already been installed and you detect damage along an edge or joint, cut out the damaged area back to sound board and prefill with joint compound. If this produces too large an area, install a filler strip of good gypsum drywall either laminated to a board layer beneath or attached with screws to the framing. Prefill around the strip and finish the joints as usual.

3.4.3 Water Damage

When gypsum drywall becomes wet, the core becomes soft and is easily deformed. Also, the facing paper may come unbonded (blistered) from the core.

If a board has been exposed to water, let it dry thoroughly before using it. Be very sure it is completely dry before installing it on the framing. If it is so badly warped that even screw attaching will not straighten it, let it dry completely and put it under a stack of new boards lying flat on the floor.

If a board is already installed and then becomes so wet that it warps away from the framing, drive in some additional screws to hold it. If additional screws do not help, take that board off the framing and replace it.

3.4.4 Board Bowing

Board bowing is similar to the warping problem previously discussed. In this case, a board may have been forced into too small a space on the framing, causing the board to bow or warp.

Whenever you discover this problem, the best remedy is to trim the board edges to relieve the stress that caused the bowing. You may have to remove the board to do a proper trim job on the edges. Reattach the board when it has been trimmed to fit properly, so that you do not have to force or pry it into place.

3.4.5 Board Cracks and Fractures

A gypsum board can crack along its face, or it may fracture all the way through to the other side. There are various causes and cures for cracks and fractures.

Board cracks may occur along the face of any drywall board, but they are most likely to show up over a doorway, where there is a smaller and weaker section of board. If a crack is more than $1/8$" wide, treat it just as you would a regular joint. Repair it by taping and feathering the joint compound and topping compound until the crack does not show.

This type of cracking is often caused by movement or settling of the building. Many larger buildings, such as skyscrapers, have a built-in flexibility that may contribute to the cracking of interior drywall. In a building with a flexible frame, the best choice is nonbearing interior partitions that have a clearance at the top of every wall. The tracks are fastened to the ceiling to hold the tops of metal studs, which may or may not be fastened to the track.

There can be up to $1/2$" clearance between the wallboards and the ceiling. Fill this space with caulk or a specialty gasket or trim such as the type shown in *Figure 40*. A control joint or expansion joint (*Figure 41*) might also be used. Place such metal or plastic trim around the appropriate board edges to give a finished appearance to the room and add protection to the walls. Any repairs you make must maintain the integrity of the wall; for example, its fire-resistance or sound rating.

Any of the following three possible causes can contribute to gypsum board fractures:

- The board was attached across the wide face of the structural framing members, such as the headers. If the framing is wood and the lumber shrinks, the board is compressed and it will crack. If the framing is steel and not adequate, loads put on it may stress and crack some of the boards attached in this way.
- The wallboard was improperly handled or stored.
- The face paper was scored past the edge of a cutout.

> **Manufacturing Defects**
>
> If you suspect that the drywall panels you are installing are defective, immediately stop work and get instructions from your supervisor. Suppliers and manufacturers will usually replace defective material. There is no sense in putting up defective material only to rip it out later.

Figure 40 Veneer L-trim casing bead.

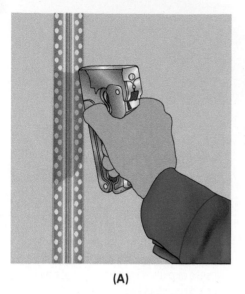

(A)

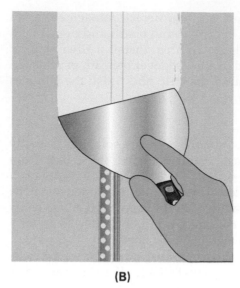

(B)

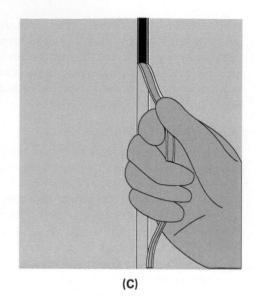

(C)

Figure 41 Applying an expansion joint.

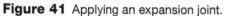

To repair broken or fractured boards, completely cut out the damaged sections and replace them. If the damage was produced by scoring the facing paper beyond the cutout edges, simply repair this score with tape as you would any other joint.

3.4.6 Loose Boards

Loose gypsum drywall boards might be caused by any of the following:

- The boards were improperly fastened.
- The framing members were misaligned, uneven, or warped (in the case of wood).
- The screws or nails were not driven in all the way or else (with lumber) some shrinkage has occurred, pulling the framing away from the board, and thereby making it loose.

Improper fastening may be due to using incorrect types of screws or an improperly adjusted screw gun. This can result in screws being stripped or not seated properly, contributing to board looseness.

The remedy for fastener problems is generally to remove all faulty fasteners. Replace them with correct fasteners and properly drive them all the way in, so that they are well fixed into the framing and produce a good dimple in the surface.

When re-driving fasteners, make sure your free hand is pushing solidly against the board near the fastening point. It is important that the board be perfectly flat against the framing member while you are driving the fastener.

Double-check to make sure you are using the correct type of drywall screw. Also, readjust the clutch on your screw gun to give you the proper depth into the board. You do not want to tear the face paper, but you need a dimple of about $\frac{1}{16}$" to allow proper spotting and finishing. If nails are used, use the double-nailing method.

If the cause of loose boards is poor framing, which may be out of alignment, twisted, or warped, your re-driven fasteners alone may not pull the board flush to where it should be. The only way to fix the problem may be to remove all the boards and correct the framing.

Another way to make a better board attachment is to use adhesive as well as additional screws to hold the board to the framing. However, if the framing is badly warped and you succeed in firmly fixing the board, your wall or ceiling might be just as warped as the framing. It is better to fix the framing.

One other possibility is to laminate an entire new layer of gypsum drywall over the warped layer using adhesive or other material. Not only will this provide new drywall laminate, but it will also fill in any spaces caused by the first layer's warping.

Finally, and most easily, if loose boards are caused by loose nails or screws, drive them in farther. Check this before finishing any wall or ceiling. Pushing with your hands against the board (even while you are spotting with joint compound) will indicate if any board is loose. If it is, stop and re-drive the fasteners. You can often drive them by simply using the butt end of your broad knife. Add other fasteners, if necessary, then continue spotting and finishing. The time you take to interrupt your finishing and fix the board-hanging problem will prevent you or anyone else from having to do the job over again.

As in all repairs, you want the fix to stay fixed. Do not settle for shortcut methods. If you have to rip off the boards and reset the framing, it is better to do it now than to have the problem reported later. If the general contractor or customer discovers the poor framing, it could cost your employer their business and could cost you your job. Fix these problems right from the start.

3.4.7 Patching Drywall

Drywall defects such as holes and dents require patching. For holes 2" or less in diameter, apply joint compound and reinforcing tape over the hole. An additional tape layer may also be needed. Once the bedding coat and tape have dried, apply a topping coat, feathering the edges. Apply a finish coat, if necessary.

Large holes require a different method of repair. One method involves using a piece of drywall. Using this method, you would square off and cut out the defective area. Bevel the edges of the squared opening so that the bevels face you. Measure and cut out a patch of new wallboard to fit this opening. Bevel the patch edges to mate with the opening's edges. Use joint compound to cement the patch in place, then tape and finish the edges as you would normal butt joints.

There are also commercially available patching systems that use fiberglass or aluminum mesh. These patches can generally be used to repair holes up to 4".

For holes 12" or larger, square off and cut out a whole wallboard section back to the framing members (*Figure 42*). Cut a fresh patch to fit, cement the patch in place, and use fasteners through the patch into the framing members. Only one drywall screw in each corner of the patch should be necessary. Tape and finish the patch edges like butt joints.

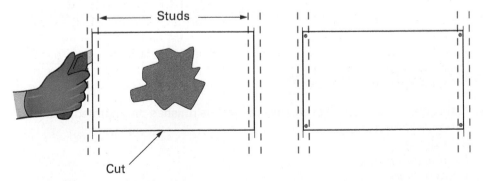

Figure 42 Patching a large hole.

To repair a wallboard dent, first sand the dented section. This raises the nap, but it also permits the joint compound to grip the drywall face paper. Fill the dent with one or more layers of compound. Allow each layer to dry before lightly sanding and then applying the next layer. Finally, sand the filled dent smooth and level with the surrounding wallboard.

Another patching technique is the hot patch or blowout patch, as shown in *Figure 43*.

NOTE

Patches must meet the UL and local code requirements for fire resistance.

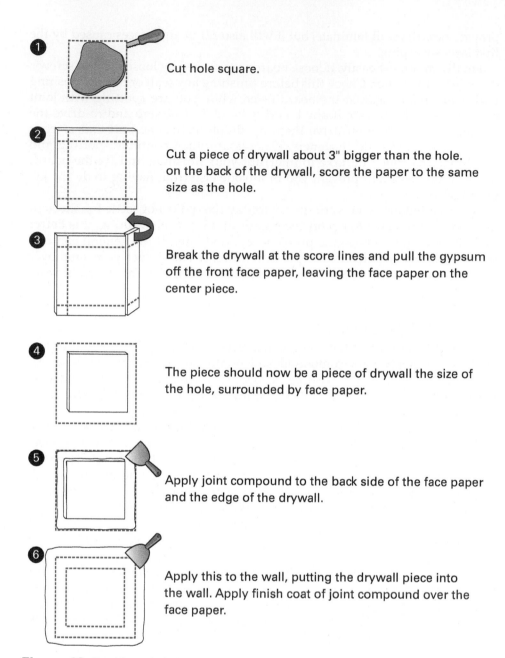

Figure 43 Patching a hole in drywall using the hot patch (blowout) patching technique.

1 Cut hole square.

2 Cut a piece of drywall about 3" bigger than the hole. on the back of the drywall, score the paper to the same size as the hole.

3 Break the drywall at the score lines and pull the gypsum off the front face paper, leaving the face paper on the center piece.

4 The piece should now be a piece of drywall the size of the hole, surrounded by face paper.

5 Apply joint compound to the back side of the face paper and the edge of the drywall.

6 Apply this to the wall, putting the drywall piece into the wall. Apply finish coat of joint compound over the face paper.

3.0.0 Section Review

1. If the joint is still visible even after the wall is finished and painted, this condition is known as _____.
   a. ridging
   b. photographing
   c. depressions
   d. high joints

2. _____ is caused when too much time elapses before the correct amount of shrinkage occurs.
   a. Sagging
   b. Excessive shrinkage
   c. Pitting
   d. Delayed shrinkage

3. When joint compound over a nail or screw has sunk lower than the surface of the surrounding drywall, it is called _____.
   a. sagging
   b. nail pop
   c. fastener depression
   d. excessive shrinkage

4. When facing paper becomes unbonded from the surface of a piece of gypsum board, it is known as _____.
   a. water damage
   b. sagging
   c. board blowing
   d. board blisters

Module 45105   Review Questions

1. Drywall tools must be cleaned and maintained _____.
   a. when they look dirty
   b. according to the manufacturer's directions
   c. once a week
   d. before you start a new job

2. A _____ is commonly used to cut drywall.
   a. hacksaw
   b. bandsaw
   c. circular saw
   d. utility knife

3. A drywall saw is good for _____.
   a. making straight cuts
   b. making curved cuts
   c. making cuts on the long edge
   d. making straight and curved cuts

4. Which of these tools is used by striking it with a rubber mallet?
   a. Finishing trowel
   b. Corner clinching tool
   c. Pole sander
   d. Mud masher

5. The automatic tool that applies a smooth finish with feathered edges and a center crown to a taped seam is the _____.
   a. flat applicator
   b. banjo
   c. flat finisher
   d. nail spotter

6. Fiberglass mesh tape may be preferred over paper tape in _____ applications.
   a. high-moisture
   b. low-humidity
   c. high-temperature
   d. low-temperature

7. You would be likely to use metal edge tape in each of these locations *except* _____.
   a. outside angled corners
   b. a 90-degree inside corner
   c. the intersection of a ceiling and a radius wall
   d. an arch

8. Setting-type compounds _____.
   a. typically take 24 hours or more to fully dry
   b. contain chemicals to cause them to harden quickly
   c. are convenient because they sand easily
   d. will have more shrinkage than drying-type compounds

9. Because it dries soft and smooth and is easy to sand, _____ is used for the second and third coats of the finishing process.
   a. taping compound
   b. all-purpose compound
   c. topping compound
   d. powder compound

10. When mixing dry-mix compounds, you are required to wear _____.
    a. gloves
    b. a respirator
    c. a hair net
    d. protective coveralls

11. Which type of texture material has good solution time, minimum-to-moderate fallout, and good bonding power?
    a. Premixed textures
    b. Powder joint compound textures
    c. Aggregated powder textures
    d. Unaggregated powder textures

12. A drywall finishing job with tape embedded in joint compound, one separate coat of compound on joints and interior angles, and two separate coats of compound over fastener heads and accessories meets the _____ requirements.
    a. Level 1
    b. Level 2
    c. Level 3
    d. Level 4

13. Drywall fasteners should be installed so that _____.
    a. the head penetrates the paper
    b. there is a slight depression in the drywall
    c. the head protrudes $1/64$" from the drywall surface
    d. the head is exactly flush with the drywall surface

14. If you discover a loose fastener during the inspection process, you should _____.
    a. notify your supervisor
    b. remove the loose fastener
    c. apply adhesive to the area around the loose fastener
    d. install an extra fastener above and below the loose fastener

15. During the final step in the finishing process, the drywall mechanic needs to _____.
    a. top the outside corners
    b. make double-wide topping coats at all butt joints
    c. complete pole and hand sanding
    d. apply a straddle coat on the butt joints

16. Which of these is the correct reason for applying multiple coats of compound?
    a. The compound shrinks as it dries, leaving depressions.
    b. A lot of the compound will flake and fall off as it dries.
    c. Extra buildup is needed to compensate for sanding.
    d. Walls look better when the seams are slightly higher than the wall surface.

17. A nail spotter allows you to _____.
    a. fill rows of fastener head depressions with joint compound
    b. countersink nail and screw heads
    c. apply tape to a drywall joint
    d. remove excess compound from depressions and joints

18. If you find bubbles in tape joints after using the automatic taper, it probably means that _____.
    a. only one wheel was pressing on the drywall surface
    b. neither wheel was pressing on the drywall surface
    c. both wheels were pressing on the drywall surface
    d. the taper was upside down

19. You should apply joint reinforcing tape _____.
    a. after the joint compound has dried
    b. where the drywall meets the floor
    c. while the joint compound is still wet
    d. after sanding the joint compound

20. Fasteners normally receive _____ coat(s) of topping.
    a. one
    b. two
    c. three
    d. four

21. One of the most common hazards when finishing drywall is _____.
    a. walking on stilts
    b. slippery conditions
    c. keeping tools clean
    d. using the correct materials

22. When you can still see a taped joint after the wall has been painted, it is known as _____.
    a. ridging
    b. photographing
    c. high joints
    d. discoloration

23. When mixing joint compound, you should use _____ water in order to avoid bonding problems.

   a. ice
   b. distilled
   c. cold, clean
   d. hot

24. To reduce the risks of nail pops, a drywall mechanic should _____.

   a. make sure there is still moisture in the framing lumber
   b. always work from the outer edge toward the center when driving fasteners
   c. use drywall to cover any misaligned framing
   d. always work from the center toward the outer edge when driving fasteners

25. To patch a drywall hole that is _____, square off and cut out a whole wallboard section back to the framing member.

   a. 12" or larger
   b. 2" or less
   c. 4" to 6"
   d. 5" to 8"

Answers to Odd-Numbered Module Review Questions are found in *Appendix A.*

Answers to Section Review Questions

Answer	Section	Objective
Section One		
1. b	1.1.4	1a
2. c	1.2.1	1b
Section Two		
1. d	2.1.0	2a
2. a	2.2.0	2b
3. b	2.3.6	2c
4. b	2.4.1	2d
Section Three		
1. b	3.1.2	3a
2. d	3.2.6	3b
3. c	3.3.3	3c
4. d	3.4.1	3d

APPENDIX A Odd-Numbered Module Review Answers

MODULE 01 (45101)

Answer	Section
1. a	1.1.0
3. d	1.2.2
5. c	1.2.2
7. c	2.0.0
9. c	2.1.0
11. b	2.1.2
13. d	3.1.0
15. c	3.1.2
17. b	4.1.2
19. a	4.2.5

MODULE 02 (45102)

Answer	Section
1. b	1.1.0
3. a	1.1.2
5. d	1.2.1
7. d	1.2.1
9. a	2.1.3
11. c	2.2.1
13. c	2.2.6
15. a	2.3.2
17. a	3.1.1
19. b	3.2.2
21. d	3.2.6
23. a	4.0.0
25. b	4.2.0
27. d	4.2.2
29. c	4.3.1

MODULE 03 (45103)

Answer	Section
1. a	1.0.0
3. c	1.1.1
5. b	1.2.0
7. a	1.2.2
9. c	2.0.0; *Table 5*
11. d	2.1.2
13. a	2.4.1
15. b	2.4.3
17. a	3.0.0
19. c	3.1.1
21. a	3.2.1
23. b	3.3.0
25. b	4.0.0
27. c	4.2.0
29. c	4.2.1; *Table 6*
31. d	5.1.0
33. b	5.3.2
35. a	5.3.2

MODULE 04 (45104)

Answer	Section
1. d	1.1.0
3. b	1.1.1
5. d	1.2.1
7. a	1.2.2
9. c	1.3.0
11. b	2.1.0
13. c	2.1.4
15. a	2.2.0
17. d	2.4.0
19. b	2.4.2

MODULE 05 (45105)

Answer	Section
1. b	1.1.1
3. d	1.1.2
5. c	1.1.5
7. b	1.2.1
9. c	1.2.2
11. c	1.2.4
13. b	2.1.1
15. c	2.2.0
17. a	2.3.3
19. c	2.4.0
21. b	2.4.5
23. c	3.2.1
25. a	3.4.7

GLOSSARY

Abuse Resistant (AR) gypsum panels: Gypsum panels designed to withstand minor abrasions and impacts against a wall.

Acoustical caulk: Caulk used to seal gaps on the perimeter of sound control walls, floors, and ceilings.

Acoustical gypsum board: Gypsum board that is designed to provide enhanced sound control and soundproofing.

Air barrier: One or more materials joined together continuously to prevent or restrict the passage of air through a building's thermal envelope and assemblies.

All-purpose compound: Combines the features of taping and topping compounds. It does not bond as well as taping compound, but finishes better.

Annular nails: Nails with rings around the shank, which provide a stronger grip and higher withdrawal resistance than nails with smooth shanks.

Apprenticeship: A drywall apprenticeship is focused on gaining on-the-job experience from those who have mastered the craft. Knowledge gained in the classroom is designed to help the apprentice better understand the job's required skills.

Batt: A flat, pre-cut piece of insulation.

Bead: An application of adhesive or other construction material in a sphere or line not less than $3/8$" in diameter.

Blocking: A wood block used as a filler piece and support member between framing members.

Box header: A type of header that is made by combining a structural panel with the framing.

Calcination: The process of heating gypsum rock enough to evaporate most of the water in its molecular structure, causing a chemical change in the material.

Cantilever: A beam, truss, or floor that extends beyond the last point of support.

Career: An occupation that offers individuals a lifelong opportunity for training, growth, and advancement.

Casing: A type of drywall trim that is used around windows and doors.

Cavity insulation: Insulating materials that are located between framing members.

Cement board: A type of durable tile backer board made of cement and reinforcing fibers.

Centerline cracking: A crack in a finished drywall joint that can occur as the result of environmental conditions or poor workmanship.

Chronic Obstructive Pulmonary Disease (COPD): Lung diseases that result in the obstruction of lung airflow and interfere with normal breathing. Chronic bronchitis and emphysema generally fall under the diagnosis of COPD.

Climate zone: A geographical region based on climate criteria, as determined by the *IECC*®.

Cold-Formed Steel (CFS): A type of steel made of sheet steel in a process that doesn't involve heat.

Collated magazine: An attachment for a screw or nail gun that automatically feeds collated fasteners (fasteners arranged side by side on a strip of plastic) into the chamber of the gun for quick application.

Combustible: Capable of easily igniting and rapidly burning; used to describe a fuel with a flash point at or above 100°F.

Competency-based training: Training that places an emphasis on ensuring trainees have the knowledge and skills needed to perform and/or demonstrate specific tasks.

Condensation: The process by which a vapor is converted to a liquid, such as the conversion of the moisture in air to water.

Confined space: A work area large enough for a person to work in, but with limited means of entry and exit and not designed for continuous occupancy. Crawl spaces and attics are examples of confined spaces.

Continuous insulation: Insulation that runs over a building's structural members seamlessly without breaks or gaps.

Control joints: Deliberate gaps left between long stretches of gypsum panels to allow them to expand, contract, or shift without cracking.

Convection: The movement of heat that either occurs naturally due to temperature differences or is forced by a fan or pump.

Corner bead: A metal or plastic angle used to protect and finish outside corners where drywall panels meet.

Corrugated: Material formed with parallel ridges or grooves.

Countersink: To drive a nail or screw through a gypsum board and into the framing member until the head of the fastener rests just below the surface, without breaking the face paper.

Cripple stud: In wall framing, a short framing stud that fills the space between the header and the top plate or between the sill and the soleplate.

Cupped-head nails: Nails with a concave head (shaped like a cup) and a thin rim.

Decoration: The application of the final surface covering on gypsum board walls.

Deflection track: A track used for CFS studs typically in interior walls to deflect the roof or floor load above without transferring axial load to the studs.

Dew point: The temperature at which air becomes oversaturated with moisture and the moisture condenses.

Diffusion: The movement, often contrary to gravity, of molecules of gas in all directions, causing them to intermingle.

Dimensional lumber: Any lumber within a range of 2" to 5" thick and up to 12" wide.

Double top plate: A length of lumber laid horizontally over the top plate of a wall to add strength to the wall.

Drywall stud: A type of CFS stud that is between 18 and 33 mils and is meant for nonbearing walls.

Edge: The paper-bound edge of a gypsum board as manufactured.

End: The side of a gypsum board perpendicular to the paper-bound edge. The gypsum core is always exposed.

Experience Modification Rate (EMR): A rating used to determine surcharge or credit to workers' compensation premiums based on a company's accident experience and potential for future losses.

Exterior Insulation Finish System (EIFS): Nonstructural, nonbearing, exterior wall cladding system that consists of an insulation board attached adhesively or mechanically, or both, to the substrate (*IBC*® 2021).

Face paper: The paper bonded to the surface of gypsum board during the manufacturing process.

Fastener pops: Protrusions of nails or screws above the surface of a gypsum board, usually caused by shrinkage of wood framing or by incorrect board installation.

Feathering: Tapering joint compound at the edges of a drywall joint to provide a uniform finish.

Fiberglass: A material made of sand and recycled glass.

Fiberglass insulation: A type of batt, roll, or loose-fill insulation that is made of extremely small pieces (or fibers) of glass.

Field: The inner area of a gypsum panel.

Finishing: The application of joint tape, joint compound, corner bead, and primer or sealer onto gypsum board in preparation for the final decoration of the surface.

Fire rating: A classification indicating in time (hours) the ability of a structure or component to withstand fire conditions.

Fire resistance: The ability of materials to prevent or retard the passage of excessive heat, hot gases, or flames.

Fire-resistance rating: The time it takes for a building material or assembly to confine a fire and continue to perform its structural function.

Fire-resistance-rated assembly: Construction built with certain materials in a certain configuration that has been shown through testing to restrict the spread of fire.

Fire safing: A type of firestopping that involves using an insulation material (generally mineral wool insulation) as a firestop.

Firestop: A piece of lumber or fire-resistant material installed in an opening to prevent the passage of fire.

Firestopping: The process of blocking openings in walls, ceilings, and floors with materials or a mechanical device to prevent the passage of fire and smoke.

Flange: The rim of an accessory used to attach it to another object or surface.

Flashing: Thin, water-resistant material that prevents water seepage into a building and directs the flow of moisture in walls.

Flexible insulation: A type of insulation that is made from a flexible material.

Floating angle method: A drywall installation technique used with wood stud framing in which no fasteners are used where ceiling and wall panels intersect in order to allow for structural stresses.

Focus four: The four leading causes of death in construction work: falls, struck-by hazards, caught-in or caught-between hazards, and electrical hazards.

Footing: The foundation for a column or the enlargement placed at the bottom of a foundation wall to distribute the weight of the structure.

Fur out: To attach furring strips or furring channels to masonry walls, wood framing, or steel framing to create a level surface before applying gypsum board.

Furring channel: A long, narrow piece of metal bent into the shape of a hat (which is why it is also called a *hat channel*), with two flanges (the brim of the hat) on either side of a channel (the crown of the hat), used to create a level surface on uneven masonry or metal framing in preparation for installing gypsum board.

Furring strip: A flat, narrow piece of wood attached to wood framing to create a level surface in preparation for installing gypsum board.

Furring: Strips of wood or metal applied to a wall or other surface to make it level, form an air space, and/or provide a fastening surface for finish covering.

Girder: The main steel or wood supporting beam for a structure.

Glass mat gypsum board: A type of gypsum board with a moisture-resistant core and face that allows it to act as a water barrier in interior and exterior applications.

Green concrete: Concrete that has hardened but has not yet gained its full structural strength.

Gypsum: A chalky type of rock that serves as the basic ingredient of plaster and gypsum board.

Gypsum board: A board with a gypsum core and paper facings. It is a building material generally used for walls and ceilings in residential and commercial buildings. It is also commonly referred to as *drywall*.

Gypsum panel: A panel with a gypsum core and glass mat facings.

Gypsum shaftliner: A type of heavy-duty fire- and moisture-resistant gypsum core panel that is used in shafts, stairwells, and chutes.

Hazard: Something that may be present on the jobsite that can cause immediate harm.

Header: A horizontal member that supports the load over an opening such as a door or window. Also known as a *lintel*.

Heat conduction: The process by which heat is transferred through a material, which is caused by a difference in temperature between two areas.

Impact Insulation Class (IIC): A sound rating that measures impact-generated noise created through the floor, ceiling, or roof that passes into the area below.

Impact Resistant (IR) gypsum panels: Gypsum panels designed to withstand stronger or more frequent impacts against a wall than AR panels.

Inside corners: Locations where two walls meet and face each other.

Job Hazard Analysis (JHA): An approach that emphasizes job tasks to identify hazards before they cause any harm. The focus is on the relationship between the worker, the task, the tools, and the work environment.

Joint: The place where two pieces of material meet. For example, the space between two drywall panels.

Joint compound: A mixture of gypsum, clay, and resin applied wet during the finishing process to the taped joints between gypsum boards to cover the fasteners,

and to the corner bead and accessories to create the illusion of a smooth unbroken surface. Sometimes called *mud* or *taping compound*.

Joint tape: Wide tape applied to the joints between gypsum boards and then covered with joint compound during the finishing process.

Joints: Places where two pieces of wallboard meet.

Joists: Equally spaced framing members that support floors and ceilings.

King stud: The full-length stud next to the trimmer stud in a wall opening.

Kraft-faced insulation: Insulation that has a paper, vinyl, or foil vapor retarder on one side.

Lath: Thin, narrow strips of wood used as a base for plaster.

Lightweight compound: An all-purpose compound having less weight than standard compounds.

Loose-fill insulation: Insulation that comes in the form of loose material in bags or bales.

Magnesium oxide (MgO) boards: A type of construction board that can be installed for a variety of uses, such as interior wall or ceiling panels or exterior sheathing, siding, or trim.

Material takeoff: A list of building materials obtained by analyzing the project drawings (also known as a *takeoff*).

Millwork: Various types of manufactured wood products such as doors, windows, and moldings.

Mineral wool insulation: A type of batt or loose-fill insulation that is made of natural stone fibers or slag.

Moderate contact: The contact between the edges and ends of abutting gypsum boards in a wall or ceiling assembly, which should not be tight or too widely spaced.

Mold or moisture-resistant gypsum board: A type of water-resistant gypsum board that is marketed as either mold-resistant or moisture-resistant, or both.

Movement joint: A type of joint that allows a building to move or relieve movement when weather or temperature changes cause any type of structural movement.

Multi-ply construction: A wall or ceiling installation built with more than one layer of gypsum board.

Non-rated assembly: A ceiling or wall assembly that does not exhibit fire-resistant properties.

Non-structural stud: A nonbearing stud for walls with gypsum sheathing.

On Center (OC): The distance between the center of one framing member or fastener to the center of an adjacent framing member or fastener.

Oriented Strand Board (OSB): Panels made from layers of wood strands bonded together.

Outdoor/Indoor Transmission Class (OITC): A sound rating that measures how well the building assemblies reduce airborne noise traveling from outside to the inside of a building.

Outside corners: Locations where two walls meet and face away from each other.

Parallel installation: Applying gypsum board so that the edges are parallel to the framing members, meaning the board is oriented vertically.

Perimeter relief: A gap left between a ceiling assembly and a wall assembly to keep the two assemblies separate and to allow the gypsum board in the ceiling assembly to move freely.

Perimeter: The outer boundary of a gypsum panel.

Perm: The measure of water vapor permeability. It equals the number of grains of water vapor passing through a 1 ft^2 piece of material per hour, per inch of mercury difference in vapor pressure.

Permeability: The measure of a material's capacity to allow the passage of liquids or gases.

Permeance: The ratio of water vapor flow to the vapor pressure difference between two surfaces.

Perpendicular installation: Applying gypsum board so that the edges are at right angles to the framing members, meaning the board is oriented horizontally.

Pitch: The number of threads per inch on a screw shank, with more threads producing a finer pitch and fewer threads producing a coarser pitch.

Plaster: A compound consisting of lime, sand, and water used to cover walls and ceilings.

Plastic concrete: Concrete in a liquid or semi-liquid workable state.

Plenum: A sealed chamber for moving air under slight pressure at the inlet or outlet of an air conditioning system. In some commercial buildings, the space above a suspended ceiling often acts as a return air plenum.

Plywood: A building material made of thin layers of wood that is used for sheathing and siding.

Radiation: Energy emitted from a source in electromagnetic waves or subatomic particles.

Rafter: A sloping structural member of a roof frame to which sheathing is attached.

Reflective insulation: A type of insulation made of outer layers of aluminum foil bonded to inner layers of various materials.

Resilient furring channel: A furring channel with one flange to attach to another surface, leaving the other side of the channel unattached.

Ribband: A 1 × 4 nailed to ceiling joists to prevent twisting and bowing of the joists.

Ridges: Slight protrusions in the center of a finished drywall joint that are usually caused by insufficient drying time. Also known as *beads*.

Ridging: A defect in finished gypsum board drywall caused by environmental or workmanship issues.

Rigid or semi-rigid insulation: A type of insulation that comes in formed boards made of mineral fibers.

R-value: A measure of an object or material's amount of thermal resistance.

Self-drilling screw: A screw with a point shaped like a drill bit to penetrate steel.

Self-piercing screw: A screw with a point sharp enough to penetrate steel.

Self-tapping screw: A screw that bores an internal screw thread in the material into which it is driven, creating a strong hold between the material and the screw.

Shaft wall: A system of nonbearing, fire-rated partitions used to enclose shafts or stairs that need fire and air resistance.

Sheathing: The sheet material or boards used to close in walls and roofs.

Sheathing braced design: A type of CFS stud bracing where the studs are braced by a sheathing material attached to one or both sides of the stud.

Shiplap: Lumber with edges that are shaped to overlap adjoining pieces.

Silicosis: A serious lung disease resulting from the inhalation of crystalline silica particles.

Sill plate: A horizontal timber that supports the framework of a building. It forms the transition between the foundation and the frame.

Single-ply construction: A wall or ceiling installation built with one layer of gypsum board.

Skim coat: A thin coat of joint or topping compound that is applied over the entire drywall surface. Sometimes required under a high gloss finish.

Slap stud: The last stud of an intersection wall, typically a T-intersection or inside corner at an intersection.

Soffit: Assembly that closes off the underside of the element of a building such as roof overhangs or beams.

Sound Attenuation Batts (SABs): Flexible fiberglass insulation batts designed to control noise in metal stud wall cavities in interior partitions.

Sound Transmission Class (STC): A rating that describes how much sound a product or assembly prevents from getting through to the other side. The higher the STC, the more sound at common frequencies has been prevented from traveling through a product or assembly to an adjacent space.

Sound-rated assembly: Construction built with certain materials in a certain configuration that has been shown through testing to obtain specific acoustical performance.

Steel braced design: A type of CFS stud bracing that involves bracing the studs using steel, whether through strapping, blocking, or running it through punchouts in the studs.

Stringer: The support member at the sides of a staircase; also, a timber used to support formwork for a concrete floor.

Strongback: An L-shaped arrangement of lumber used to support ceiling joists and keep them in alignment. In concrete work, it represents the upright support for a form.

Structural stud: Extra heavy CFS or wood stud that is used in the exterior or structural frame of a building, is load-bearing, and is used to resist environmental loads.

Stucco: A type of plaster used to coat exterior walls.

Studs: The vertical support members for walls.

Tape: A strong paper or fiberglass tape used to cover the joint between two sheets of drywall.

Tapered joint: A joint where tapered edges of drywall meet.

Thermal bridging: When a small area of floor, wall, or roof loses substantially more heat than the surrounding area.

Thermal resistance: A measure of how a material will resist the flow of heat energy.

Thread: The protruding rib of a screw that winds in a helix down its shank.

Tile backer board: A board that is used to underlay ceramic tiles in wet areas of a building.

Tooth: A textured surface created mechanically to help a covering material adhere more effectively.

Top plate: The upper horizontal member of a wall or partition frame.

Topping compound: A joint compound used for second and third coats. It dries soft and smooth and is easier to sand than taping compound.

Track: A length of steel that goes on the top and bottom of the wall or ceiling to receive steel stud framing members.

Trimmer stud: The vertical framing member that forms the sides of a rough opening for a door or window. It provides stiffening for the frame and supports the weight of the header.

Truss: An engineered assembly made of wood or metal that is used in place of individual structural members such as the joists and rafters used to support floors and roofs.

Type C: A type of interior gypsum board that is rated as fire resistant and has more glass fibers than Type X, as well as unexpanded vermiculite components.

Type X: A type of interior gypsum board that is rated fire resistant and has a gypsum core with the inclusion of glass fiber strands.

U-factor: A measure of the total heat transmission through a wall, roof, or floor of a structure.

Underlayment: A material such as plywood or particleboard that is installed on top of a subfloor to provide a smooth surface for the finish flooring.

Unfaced insulation: Insulation that does not have a vapor retarder.

Vapor permeable: Permitting the passage of moisture vapor. A vapor permeable material, as defined by the *IBC*®, has a moisture vapor permeance of 5 perms or greater.

Vapor retarder: A material used to retard the flow of vapor and moisture into walls and prevent condensation within them. The vapor retarder must be located on the warm side of the wall.

Vapor retarder class: A measure of a material's ability to limit the amount of moisture that passes through it.

Vaulted ceiling: A high, open ceiling that generally follows the roof pitch.

Veneer: The covering layer of material for a wall or the facing materials applied to a substrate.

Water vapor: Water in a vapor (gas) form, especially when below the boiling point and diffused in the atmosphere.

Water-resistive barrier: One or more materials installed behind exterior wall coverings to prevent water from entering a building.

Withdrawal resistance: The amount of resistance of a nail or screw to being pulled out of a material into which it has been driven.

Z-furring channel: A Z-shaped steel channel that is used to furr out walls and to create a uniform surface for gypsum board.

REFERENCES

45101 Introduction to Drywall

The Gypsum Construction Handbook, Seventh Edition. USG Corporation. Hoboken: Wiley, 2014.

NCCER Construction Craft Salary Survey. 2022. https://www.nccer.org/docs/default-source/pdfs/2022_constructioncraftsalarysurvey.pdf?Status=Temp&sfvrsn=204904a_2

Occupational Safety and Health Administration (OSHA), http://www.osha.gov/law-regs.html

45102 Construction Materials and Methods

American Iron and Steel Institute (AISI). https://www.steel.org/

ASTM International. https://www.astm.org/

Gypsum Construction Handbook. Chicago, IL: United States Gypsum Company, 2000.

Steel Framing Industry Association. https://sfia.memberclicks.net/

45103 Thermal and Moisture Protection

International Building Code®, 2021. International Code Council.

International Energy Conservation Code®, 2021. International Code Council.

International Residential Code®, 2021. International Code Council.

LEED Reference Guide for Building Design and Construction, v4. US Green Building Council.

Standard 90.1-2022—Energy Standard for Sites and Buildings Except Low-Rise Residential Buildings. ANSI/ASHRAE/IES.

Standard 189.1-2020—Standard for the Design of High-Performance Green Buildings. ANSI/ASHRAE/ICC/USGBC/IES

US Department of Energy website, www.eere.energy.gov

45104 Drywall Installation

The Gypsum Construction Handbook, Seventh Edition, 2014. Chicago: USG.

45105 Drywall Finishing

The Gypsum Construction Handbook, Seventh Edition, 2014. Chicago: USG.